CONVERSATIONS WITH THE LAND

JIM VANDERPOL

FOREWORD BY DAVID KLINE

NO BULL PRESS

Copyright 2012 by No Bull Press LLC and Jim VanDerPol
All rights reserved
by No Bull Press, LLC
Brooklyn, WI

Cover design and book layout by Curtis Dorschner
Illustrations by Carye VanDerPol Mahoney: http://carye.mosaicglobe.com/
Cover photos by Joel McNair

ISBN 978-0-9839503-0-1
Library of Congress Control Number: 2011942852

Farm Life — Minnesota
Agriculture — Social aspects
Sustainable development
Social justice

No Bull Press
Brooklyn, WI
www.nobullpressonline.com

CONTENTS

Acknowledgements i
Foreword (by David Kline) ii
Introduction iv
Seasons 1
    Lilac 2
    Smell 4
    Thunderstorm 6
    Weather Report 8
    Time & Change 10
    Autumn 12
    Hunkering 14
    Christmas 16
    Dakota 18
    Winter 20
    Seasons 22
Characters 25
    King 26
    Laughing Dog 28
    Dog as Kid 30
    Dogs 32
    Truck 34
    Steel Wheels 36
    Mort 39
    Ellis 41
    Truckers 44
    Geo 47
Wives, Husbands, Children 50
    Death & Rebirth 51
    Kite 54
    Bulls 56
    Her Land 58
    Husbandry 60
    Boys 63
    Children 66
    College 68
    Adult Children 70
    Crash 72

CONTENTS

Communities 75
    Hay 76
    Decline 78
    Alice 81
    Pao Yang 83
    Local 86
    Wrong Story 88
    Goose 91

Enabling The Scoundrels 94
    Slaughterhouse 95
    Painter 98
    Superstition 100
    Deer Rifles 103
    Stunned 105
    Church & Capitalism 108
    Failure of Citizenship 111
    Money & Morality 114

How We Might Farm 117
    Jake's Fence 118
    Patterns 120
    Art & Science 124
    Elegant Solution 127
    Fundamentalism 129
    Durability 132
    Farming 135

How We Might Live 139
    Stickers & Drifters 140
    River 143
    Ready for Change 145
    Food & Health 148
    Real Change 150
    Archer 153

ACKNOWLEDGEMENTS

I have much gratitude to express to those who have helped me with my writing.

The first is LeeAnn, who has been persecuting me about a book for more than forty years, or ever since I pursued her until she caught me. She believed in me before there was anything much to believe in, an act of faith for which I am forever grateful. There is Joel McNair, my editor at Graze magazine and publisher of this collection, and his wife Ruth McNair, who is handling the nuts and bolts. I cannot forget my daughter, Carye Mahoney, who did the line drawings for this book, son-in-law Curtis Dorschner, who designed the cover and book layout, and my son Josh and his family, who helped take up the slack when I was writing instead of farming. John White, editor of the Clara City, Minnesota, Herald, came up with the "Conversations with the Land" title when I was writing for him. Thank you, John.

I am grateful to the people of the Minnesota-based Land Stewardship Project, which helped start me in the conversation. Also, Minnesota's Sustainable Farming Association impressed me many years ago with the rightness and hopefulness in the practice of farmer-to-farmer networking and the local foods movement, which is putting feet under our thoughts and words. You are fellow travelers all, and your numbers are significant and growing. You inspire me. Your creativity and dedication give me hope. I especially thank the "graziers" — the people who have resurrected the seemingly lost art of grazing livestock — and their groups who have made me feel at home with them, even though I raise hogs instead of milking cows.

FOREWORD
BY DAVID KLINE

In prose as rich as the soil of his Minnesota farm, Jim VanDerPol shares with us in this book what he and his family have done since taking over his parents' farm in 1977. We smile and ponder, laugh and learn, as we read about this family farm. Jim soon realized that the corn and soybean rotation that was becoming the accepted pattern for farming in his region would not support his family on a modestly sized farm. He didn't heed agribusiness's battle cry of "Get big or get out!" Being a prairie populist and true nonconformist, he took a stand against the huge and powerful forces working against family farming. Jim opted to become a much more diversified farmer and has succeeded remarkably well, as these essays testify.

These essays show us that agri-culture and agri-business are two entirely different worlds. Agribusiness tends to mine the land with a "what's in it for me" mentality and a human cleverness world of chemicals and gadgets — a life similar to golf, where one spends a lot of time riding around.

Jim's obvious choice is agri-culture, where human touch, kindness, love of land and neighbors, and working with nature and the seasons matter. It is a full and generous life where one is a nurturer instead of an exploiter. He understands that human and humus spring from the same root. It is apparent that Jim and his family live on the land, rather than merely residing on it. The garden and the sauerkraut stored for winter are as important as the fields, the wildlife are as important as the livestock, and the balance sheet for the year shows a profit from the capture of contemporary sunlight instead of ancient fossil fuel.

While we all know that making a profit on the farm is crucial to staying on the land, Jim clearly points out that money in itself is not an accurate scorecard in measuring life. On the VanDerPol farm, physical labor is not drudgery. Jim and his grandsons muck out corner stalls inaccessible with the skid-steer, and Jim knows that some of the best discussions and conversations come about while forking out manure. Perhaps it is the rich and moldy aroma that sets off the conversations. We diversified farmers are fully aware that technology is a terrific servant, but a terrible master. To us, pitchforks are still useful tools.

In 1998 the VanDerPols began direct-marketing their grass-based pork and beef. Not only has it worked out well for them (their 320-acre farm now

FOREWORD

supports two families), it has also benefited local people, from the trucker to the butcher. This is a model for the farms of the future, as crude oil prices again are creeping over a hundred dollars a barrel.

I met Jim sometime in the mid-1990s when Farming Magazine had barely entered its incubation stage. When I mentioned our wild dream of a publication advocating diversified, sustainable, small-scale and profitable agriculture, he immediately said, "Count me on board. I will help in any way I can." He has and continues to do so, for which we are extremely grateful. His columns, also called "Conversations with the Land," are read and appreciated by many people in both our magazine and Graze, a magazine for people who are returning their stock to the land.

Yet quite a few of these columns are new to me, such as the one about thumbless "Geo," and also "Ellis." Ellis was Jim's neighbor and friend, and a farmer of the old school. This is what Jim writes about his friend:

Ellis thought that farmers should want to live mostly on the farm instead of at the lake or casino, and he could not, for the life of him, figure out why so many who surrounded him in his later life did not. He loved it all, from the rain that wouldn't come or wouldn't stop when it did, to the trees that scratched his tractor, to the crop that amazed him one year and disappointed him the next. He loved sweat and the tears and the loss of it, because he knew that without those drawbacks he didn't get the new life, the new shot at it the next year, the dew on the grass, the birds in the trees, the possibility of rest and the comfort of neighbors ... Ellis was buried in the cemetery next to where the country church stood, the one he attended for so many years among friends, relatives and neighbors. There are a few cottonwoods nearby for music when the wind blows, and he can keep an eye on the corn.

This is a book about good farming. Henry Thoreau, and Gilbert White before him, wrote that a man has to "see before he can say." Jim VanDerPol travels a lot at home, where he sees much, and says it very well.

CONVERSATIONS WITH THE LAND

INTRODUCTION

In 1952 my parents purchased this farm, surely one of the wettest and most difficult in western Minnesota's Chippewa County. I grew up here, took my turn at the University of Minnesota in the late '60s passing classes and playing at hippie revolution, and then married and returned to the farm with wife, young son and infant daughter in 1977.

I had decided to farm. My wife had decided to be patient, for which I am forever grateful. My children, including the last one who was born in the early years of our farming, grew up here building their own impressions and, I suppose, illusions about the world's reality. I can only surmise that those impressions are different from mine, even though they were formed in the same place.

The farm was a fairly typical corn and soybeans rotation when we started here, with just a small planting of small grains, usually wheat or oats — an idea left over from the diversified farm it once was. There were no pastures, no hayfields and no sloughs that we were willing to admit to. Even then it was too small at five hundred acres to support a family with that kind of production scheme. In our first year, we purchased three bred gilts and began the hog operation that was to define the farm for our lifetimes.

Like any farm, we had our share of disasters. There was the farm financial crisis in the mid-eighties, the drought of 1988, the hog price meltdown of 1998. Interlaced were several dire hog disease outbreaks, reproductive failure due to bad feed and several land price run-ups, including the one happening now.

In reaction to these major events, and in keeping with growth in our thinking about farming, we changed from a fully conventional farm with annual crops, first to reduced tillage through a ridge tillage system. This was in reaction to the dust and wind erosion we saw in the drought of 1988, and also as a nod to our personal preferences for less tractor work. But dissatisfied with the simplified rotation that had been reduced to corn and soybeans only, plus the difficulties of operating a wet farm with minimal tillage, we moved toward perennial crops in the form of pasture production — first with a small sheep flock as well as the sows, and then into a full-blown grazing cattle business.

INTRODUCTION

In making this move we were much influenced by the thinking of Alan Savory about grazing and desertification. This was a time of big change for us, as our son and daughter-in-law also joined us on the farm. The farm is now forty percent in permanent grass pastures producing cattle and maintaining the sow herd, and sixty percent in a cropping scheme that includes perennial hay crops as a part of the rotation. And we do have a fifteen-acre slough that we are slowly building into a large farm pond, working at it whenever it is dry enough and we can afford to hire a bulldozer.

The move into pork marketing followed the market-price disaster of 1998. The business grew, and we learned more about the marketing of a farm's products. As luck would have it, we were starting the effort at about the time the cooperative grocery stores and high-end markets in the Twin Cities were looking for a local producer of pork willing to use humane production protocols.

In the background of much of this was the willingness of my wife to work off-farm to provide the family-living income. Nothing can be said or is going to be said by me to discourage one spouse from working off the farm. But it is a fact that while this working off-farm was an economic necessity for us, the time driving to and from that work might better have been spent with business and family. Even though the work involved was good and necessary, it was a distraction and a hardship for the one doing it, for her other responsibilities never let up. This farm would not exist today were it not for the twenty years LeeAnn spent at that extra job.

I was born in the late 1940s into an agriculture community that was failing. In the manner of a child, I assumed the world I was part of was permanent and ongoing. My living and farming and writing since that time can all be seen in terms of a dawning realization that this was not so. Indeed, I soon saw that not only was change happening, but that something valuable was being carelessly tossed aside. The knowledge of how to farm, as well as a passion for the occupation, was being lost each time a farm failed. I saw that I was becoming alone, isolated and an outlier in my home country.

I became angry. Anger by itself is a corrosive and destructive emotion. But anger that has an object — anger that knows it has been messed with and by whom — is wholesome, character-building and very effective as an agent for change. Many of the essays in this book are built upon that anger. I think they are some of the best. Some of them point to huge forces that mean to suck the economic life from us, and some are directed at what I have begun to call "the man in the mirror." This is because many of the most vicious attacks by the powers-that-be upon our economic wellbeing are enabled by the mistaken attitudes we carry around.

From my awakening anger it was but a short step to the determination to do something about it. That determination, and the capacity for that anger, was put into me by two true prairie populists: my father, who taught me to care for family and love farming, and my mother, who taught me to care for family and love learning. They knew that success was not a given, but also that giving up was not an option.

So it was that not so many years into farming as the universities and corporate agriculture would have it that I found fellow travelers, and went the other way. Telling about it and talking about it with others seemed the natural thing to do. It became apparent that writing about a life lived inside the problem can be a powerful and potentially healing thing.

Now, from the perspective of a full twenty-five years into the changes and the conversation, it is becoming impossible to avoid the realization that the problems and misdirection I have been speaking about are not exclusively rural, that some version of the same destructive force is affecting all Americans and, quite possibly, most of the world as well. This thought is both terrifying because of the size and power of the destructive force, and exhilarating because, if the problem is really that widespread, think of the millions of people who should be ready to join the conversation!

Agriculture is a human endeavor. We have forgotten this, or have allowed our fascination with our crackpot economic philosophy to drive it from our minds. One of the goals my family set for our farm is to work surrounded by more people. Because that goal is there, we have made decisions in such a way that some of the goal has come to pass. Official agriculture, meanwhile becomes more inhuman and mechanical with each new machine and each new manufactured seed and hormone. Geopositioning results in absolutely straight corn rows, and for the time being Roundup makes them weed free. Nature does not do this kind of work. Neither do people. Not surprisingly, as the technology has taken control and the management has left, so have the people. At this point conventional agriculture cannot find people enough in the younger generation to carry the enterprise forward. It is increasingly an old man's game. Farming by riding around.

It is people, often female, who can empathize with the farrowing sow. It is people, not orbiting satellites, who through their feet can feel the painful changes in the earth when it dries up and blows around. And it is only people

INTRODUCTION

who can deal with climate change and land degradation and desertification, each in his or her own life. No institution can do much more than talk about it. Government is captive to the forces causing the problem.

But if they choose, people can take care of each other, the surrounding community and the earth on which they depend. This is what being human means. And this is why the conversation is so important. We need to teach each other how to properly live on the earth. An agriculture that conspires to drive people out of it has a death wish. This is so apparent now that it hardly bears stating. The task for all of us, breathtaking in its scope, is to try to make sure it does not destroy everything around it in the fulfillment of that wish.

It will not do to discount the impact that applying our full humanity can have on our living circumstances, no matter who we are, where we live or what we do. My youngest daughter, Katie, asked me twenty years ago as part of a school assignment why I chose my occupation. I told her that I farmed because it was something that used all of me, my mind as well as my body. That is as true today as it was then, and I might be forgiven for offering that thought as an explanation for many of our current social problems. It is that willingness to think about what we are doing while we are doing it that both improves the work and the possibility of success in doing it. For me, inclined as I am to reading and writing, column writing was a natural fit with farming. I can honestly say that most of the columns occur to me about half written while I am busy doing physical farm work. And I can also say that the writing has had an important impact on our farming. It has the effect of supplying another eye on the enterprise.

It is my hope that you, the reader, can take these essays with a grain of salt. I wish to be as direct and honest as possible. I do not believe objectivity is possible, so I offer this short description of myself and my beliefs so that you may be forewarned.

I am a liberal, or progressive, if you prefer. This means I believe in certain policies, not in a certain party. The religion in which I was raised instructs that I am to be my brother's keeper, and I have never been able to figure out how to do that in our times without involving the government. I am in step with the Democrats whenever they are useful, which unfortunately they most often are not. When they are not, I go my own way. In my personal life I am conservative, upholding as best I can the values of thrift, personal responsibility, humility and care of the land and community.

The difficulty with being politically conservative in America is that the philosophy is confused and dumbfounding when it comes to economics. Conservatives insist upon a rigid belief in capitalism, which is not conservative at all, but incredibly radical. Capitalism destroys resources and people with equanimity as it seeks to turn them into money with all possible speed. It gives no thought to the future and takes no responsibility for damage to either the natural or human community in its quest to gather all wealth to itself. And "itself" is always fewer and fewer people. Liberals, without much in the way of success, seek through government regulation to control the viciousness of capitalism while preserving some of its productive results.

Lately I have become attracted to the idea of cultural controls on capitalism, and I have written about the idea. We all know stories of smaller good businesses and farms, some of them multigenerational, that run without damage or with minimal damage to the resources they are using, the people doing the production or the customers. I wonder what it is about these people that seems to allow them to use the tools of capitalism without participating in its destructiveness. I have begun to think that exploration of this idea and phenomenon holds the most promise for fostering a good and conserving agriculture and rural community in America.

— November 2011

CONVERSATIONS
WITH
THE LAND

JIM VANDERPOL

Illustrations by Carye VanDerPol Mahoney

CONVERSATIONS WITH THE LAND

SEASONS

CONVERSATIONS WITH THE LAND

LILAC

A lilac planted just last summer sticks its leader up from the two-foot snowbank next to the house. I walk past the lilac every morning on my way to the barn. It has a half-dozen buds, still hard against the winter. I stop often to check them, knowing that a day will come when the buds will be softer, and then soon softer yet. Soon, then, I will be able to roll one of them between my fingers and get a sticky sap that I can then taste for astringency, with just a hint underneath of green leaves and a fleeting sense of floral. When that happens, the remnants of the snow bank are likely to be lying around the lilac stem, and I will celebrate that the lilac is winning over winter. It is not, of course, as like winter it is taking its turn.

It will not do to minimize the sense of celebration after this difficult winter here on the northern plains. Each year, as we pass the midpoint of February and the days become several hours longer than at Christmas, I can feel the life and strength flowing back into my body. Celebration is the only word that does it justice. Dancing a jig in chore boots does not make a pretty sight, especially if the dancer is getting old, but generally nobody is looking except the cattle and a couple of rooster pheasants that have come to find the feed the sows left yesterday. They are all blessedly nonjudgmental.

I have often remarked on the ways in which my body has become attuned to the natural world here where I decided to live so many years ago, and I cannot help but feel that it would not be the same had I decided to stay in the urban area where I was in my twenties. Of course, in this almost completely urbanized nation, there are only just a few of us who may know what I am talking about.

I am reminded of something I heard in a sermon twenty-five years ago that has since acquired a kind of double meaning for me. The preacher told the story of a government employee, a single man who had lived and done his work in Washington, D.C., for many years. Each year the man would take his vacation in a remote, rural area of Maine, staying in the spare bedroom of an elderly widow and taking in the peace and beauty of the lake and countryside surrounding the house. She had never been out of the county, he knew. One year, partly out of gratitude for the refreshment and peace offered him each year, and partly out of a sense of obligation to do what he could to broaden her perspective on life, he began on his last day there to tell her of his job and circumstance, and to describe for her what it was to live and work in the nation's capital.

After he had gone on in this way for twenty minutes or so, he ran down and they sat quietly for a few minutes. Finally she said, half to herself: "Just think of all those people living there in DC, so far away from everything."

Now that I have gotten older and have a little more perspective on what can and cannot be expected of our government or perhaps of any government, I can see in this exchange a meaning I didn't get before. This can be heard as a real and accurate criticism of urban society and the corporate elites who are so much a part of that society. One can go on to analyze how the woman was amazed at the thought of DC because it never did her any good, and yet holds so much power and influence in the world, even if it is not often used for anyone's real benefit.

But I think the meaning I saw in the first place is the important one. It has to do with living somewhere, with being home in a place on earth. What is startling about it is that from the woman's perspective, home and a good life do not require a huge number of people, or even a large amount of power or money. It requires enough gumption to see to herself at least partly, to place importance upon things the urban elites tend to ignore, and the knack of wanting what she has instead of what she doesn't have.

The woman in the story would understand about my lilac. I am sure of it. She undoubtedly looked at lilacs and other buds in the spring, tasting and smelling and dreaming over them, and picturing herself and her gardens and her animals in the long summer days to come. She, too, felt the new strength flowing through her limbs each February.

I, too, would wish for springtime for that strange, benighted crew in Washington, D.C., and other centers of power. I would wish they could learn to appreciate us for whom and what we are. They could start by counting us as people instead of mere votes. I wish they would quit taking our wealth, the birthright of our children, and handing it over to those who wish to make it into so much junk and waste. I wish they would quit mistaking our good nature for acquiescence to their continuous wars.

I wish they could see that there is no hatred in a springtime bud. I will supply the lilac branch to teach them. My farm is rich with them, and can spare a few.

CONVERSATIONS WITH THE LAND

SMELL

Many folks who work with older people believe that the sense of smell is a jog to the memory. Even those who seem lost in themselves can be seen to perk up and come to attention at a familiar smell, something as common — or perhaps uncommon in this day and age — as a pie baking or a chicken dinner. It is not that the sense of smell gets stronger with age, but that it seems to provide the surest link with what has gone before. I notice it myself.

The nose can tell you what time it is on a diversified farm. There is the damp, rich and almost moldy smell that comes from worked soil in the spring. It is possible, if you get far enough away from the tractor and the diesel fumes, to use your nose to tell whether the field is fit to be worked or planted. Too wet smells a little sour. Dry is dusty, irritating to the nose.

Then there is the composted, fermenting smell of manure and bedding when it is stirred up and hauled out. This blends nicely with the earth smell, and in a day's time you can't tell one from the other.

Each species of livestock has its own smell, and none of them are unpleasant. It angers me that animals get blamed for how livestock facilities and manure smell. In my view, engineers and banks ought to come in for the blame, because they are building the wrong stuff. Even hogs have a rich, clean smell except when it is hot, when they smell of mud. The septic smell they get blamed for has nothing to do with hogs. Anyone knows this who has ever stood in an open barn filled with sows nursing litters on straw.

Most stock smell a little different on pasture, with the smell of the animal mixed with the chlorophyll from the plants they are chewing. Some pastures, particularly annuals such as rape and turnips, get the digestion working so fast that you can smell fermenting feed from both ends of the animal.

Ruminants such as cattle and sheep always have a yeasty, "brewy" smell to them. I wouldn't be surprised if your blood alcohol level didn't bump up a little when a stocker steer on alfalfa comes up and snorts a big breath right into your face. With sheep, you also get a kind of soapy, oily smell that has to do with the lanolin in the wool.

Then there is the heavy, moist and dangerous smell that comes just before a thunderstorm hits, followed by the clear, heady ozone smell right after.

SEASONS

Between storms, as spring gives way to summer's heat, the earth gives off a kind of baking aroma, as if it is being cooked a little and becoming overdone. Plants are smelling more mature now, and the honeybees and bumblebees search out the alfalfa fields and thistle patches by the way their sweet smell contradicts the surrounding ripening odors.

This is the time of year when the corn smell on hot evenings becomes almost overpowering, and the old timers lean back on their porches and claim to be able to see the corn grow. And of course the young mutter, "Yeah right, Grandpa," mostly under their breath.

The fall is all sharpness. Allergy sufferers are suffering. The air is full of dust, and the plants are busy protecting their ripening seeds from the foraging blackbirds by all manner of bad tastes and nasty smells. The cumulative effect of all this is a kind of fullness made up of dust and fragments of plant leaves and the curing of plant material under the sun.

Winter smells empty, kind of how the land feels. There is an end-of-the-line feel to it. You sniff and try to figure it out, but all you get is a burning feeling from too much cold air. But that sharp, clean smell so common in winter will give way to the fermenting, awakening smells of spring. We know that from experience. That is why we keep an eye on the day length as we do the chores, and kind of cheer on the sun under our collective breath. Spring has not failed to come yet and, somewhere deep inside, we want to make sure it doesn't fail this time.

Come to think of it, thirty years ago it may have been my nose that led me out of the city and back to the country. The smell of bus diesel and car exhaust mixed with the acrid stink of last year's leftover road salt was not going to do it for me.

Stephen King and Peter Straub, who wrote *The Talisman* together, created a wolfboy character in the underworld whose nose was so developed he couldn't stand the smell of asphalt when he came to our side. Said it smelled of dead bodies, which of course it is.

Now that's a nose!

CONVERSATIONS WITH THE LAND

THUNDERSTORM

At about three o'clock on Sunday when the second rain started — the one that turned out to be a half inch in about ten minutes — I ducked inside the chicken house and went about feeding and watering them while listening to the rain drumming on the steel roof. After the chores were finished, I turned a water pail upside down and sat in the doorway, watching it rain.

The water was coming off the hoop house tarpaulin in a continuous sheet all along the tie-down pipe. It hit the ground and then joined the little river running from up near the barn and flowing down toward the pasture. The stream grew as it picked up more and more runoff, and soon the twigs and little mounds of soil and hay further up the hill were washed loose and carried off.

I wonder how many farmers have spent how much time sitting on upturned buckets in open barn doorways, watching it rain. Sometimes the rain is wanted and needed; sometimes it is just so much more water that gets in the way.

There is this, though. The rain is always something of a gift because it stops you and makes you take notice for a while at least. If there weren't thunderstorms, we would probably have to invent them. What do folks who don't live their lives out in the weather do? What stops them and brings them down from the rush and bustle?

A good, hard thunderstorm in the middle of haying is a blessing for a teenager. Haying always happens on the weekend. And back when haying meant plenty of physical work and sweat, it also happened on the hottest, most humid days of the summer. When we stood inside the barn looking at the rain bounce three feet back into the air after hitting the ground, and felt the super-cooled air through our sweat-soaked shirts, we knew one thing: We were going to get to town on time tonight for a change, and with any luck there would still be a few girls around.

The only thing that stood in the way was the cows. If the rain would let up pretty quick, we could get down to open the pasture gate and let them up for the milking. When the rain shuts you off from haying at five o'clock and you are seventeen and it is Saturday, milking is something that is not going to be allowed to stand in the way for long. Meanwhile, the older guys are smoking and mulling things over at the barn door. The group usually consisted of Dad, an uncle and several neighbors. While we were tripping over them getting the milking equipment ready, they were busy weighing the need for rain against the ruined hay, and wondering if the corn would make up for it. A thunderstorm is one of the few occasions I ever saw that bunch of men sitting still, other than while they were eating.

Today I weigh the ruined hay against the corn's need for rain, and don't think quite so often about who might be in town. And the others I grew up with mostly don't think of corn and hay at all in the context of rain.

If you watch the Weather Channel, you know that rain is viewed differently now. It is all right as long as it doesn't fall on the weekend and ruin recreation. And it better not slow traffic on the major freeways, which are there on the screen in plain view. While the freeways are marked on weather maps, rivers, which actually affect weather and are affected by it, are not shown except where they happen to form a state's border.

So rain is all right in any amount during the weekdays as long as it happens between ten and three so it doesn't interfere with the "commute," and as long as it doesn't flood my basement. And the river ought to stay within its banks. If those conditions are met, all is right in the Weather Channel's view.

Maybe not having to look at a freeway through the windshield wipers on the way to work in the morning is as important as a ruined hay or corn crop is to a farmer. I don't want to wish anyone ill. And I know that anyone who writes about anything that happened a few decades ago is going to be accused of sentimental nostalgia by all of our modern high priests of the bright and shiny future. Probably kids are better off in the high school weight room than they were stacking hay in a hot barn.

But maybe we have moved a bit too far away from our sources. It is to be hoped that we can find our way back when we need to.

CONVERSATIONS WITH THE LAND

WEATHER REPORT

Few things in the culture that surrounds us are as determinedly urban as the weather report. With the exception of the reports that are bent a little out of their usual shape in order to fit a country radio station's farm report, weather is always given the same urban and leisure slant.

The weather features a clean-shaven fellow dressed in a three-piece suit. Or more likely in view of our modern need for political correctness, a young lady — probably a former winner of her suburb's summer celebration queen contest — reading us the weather according to the weather service's computer. Her diction is correct, her posture beautiful and she, like her entire audience, is waiting for the weekend.

The patter is about "your morning commute" and "your drive home." Of great concern is the existence and length of airport delays due to weather at "travel destinations" which are all large urban centers. Rain ruins the view out of your office (everyone works in an office) window, but that is better than having it ruin your weekend on the beach. One wonders if any in Florida lately have connected all those beautiful, rainless beach weekends with the smoke-producing wildfires.

In the world of television weather, no one is so unlucky as to have to work or live in a place that is neither air-conditioned or heated, so the weather person's main concern during the week is that we all take our umbrellas or overcoats so that we can make it from house to car, and then car to office. Outside is just a little slice of space we get through on our way to somewhere important.

Except on the weekend. And remember, on the weekend rain would RUIN EVERYTHING! If only we could dome the earth and air condition it! Author/farmer Wendell Berry observes quite correctly that the destruction of the earth is being hatched in those minds more surely even than in the Pentagon.

There are a few of us who are still in a kind of pre-modern world where we live every day in the weather. Farmers would be one example, and by farmers I mean those who still do their own work. So do electric utility linemen, telephone company workers and road maintenance people. And any carpenters, electricians and plumbers, along with most truck delivery drivers and many mechanics, just to name a few occupations that must deal with the elements. You will notice that most of the folks on this list are pretty busy providing goods and services that allow many others to live their protected lives.

Farmers have a vital interest in the weather because it has to do with how good or bad the crops and pastures will be, and thus how well the livestock, and consequently the pocketbook, are doing. So besides living every day in the weather, we farmers more than anyone else live or die (at least economically) by the weather.

It would be helpful if more people in this country gardened, for gardening teaches many of the same kinds of things as farming: that rain can be welcome, that there can be too much sun and heat, that winter is necessary to prepare for the next season. A thunderstorm that ruins the picnic may be something to dance for joy about if the beans needed a drink badly enough.

It is perhaps inevitable that something as important to the farmer as weather be taken with a certain fatalism. It is in any case necessary, for we know more than most that we are not in control of the weather, even admitting in our saner moments that it is a good thing we are not. The crops will thrive, or they will die. In any case, whatever is going well or wrong with this year will end, and we will have to face next year. And we will likely do that with all the inner joy and optimism that we always feel for that annual new chance. We never learn. That must be part of what it is to be a farmer.

In any case, understanding that we do not control the weather, and that one way or another we will live with what we get, is miles ahead of the intellect behind that brain-dead television smile promising us a "perfect weekend" while our crops are drying up.

I have decided that I am going to pay more attention to the weather and listen a great deal less to the weather report. Anyone who is wrong as often as the weather service should probably not be accorded the respect that goes with being listened to so much. And listening to some computer's guess about tomorrow's weather may very well be dulling my own powers of observation. That would be more of a shame than rain on the weekend!

CONVERSATIONS WITH THE LAND

TIME & CHANGE

Time is different for a farmer. Of all the things I would have thought I might notice when we decided to start farming a couple of decades ago, a different idea of time would not have been one of them. But soon after we started, I stopped wearing a watch.

This began logically enough, as it was just that I couldn't keep the things on my wrist. They kept tearing off on machinery and being laid aside for livestock work and sometimes not being found back. I tried a pocket watch, but lost that riding the Tilt-a-Whirl at the county fair with the kids. Soon I got tired of replacing watches. Very soon after that I found I didn't want one anymore.

Farming is one of the few occupations that takes place largely out of doors. I know this is changing, with crop farmers spending their days in tractors and shops and in front of computers, and livestock farmers working in buildings and in front of computers. Yet much time is still spent out in the weather.

This makes sun time a real alternative. Sun time isn't as accurate as clock time. For example, I can usually show up within a half hour of noon, which is close enough if the cook is very patient or if I am cooking myself. For most purposes it is accurate enough. There are wonderful things about this. Break time is when you are hot, winded, tired. Suppertime is when you're hungry. Chore time is whenever that seems right. Not being a slave to the clock means that the whole day can be arranged to suit the mood.

Of course there are more demanding aspects as well. Quitting time is not five o'clock; it's when the work is done or the farmer plays out, whichever happens first. Start time is whenever you can get yourself awake.

This whole idea spills over into the seasons. We had not been at farming many years when I first noticed that I seemed to need less sleep in the summer. Winter is different, requiring about three more hours in the sack. It seems natural enough if you consider that the sun is acting in about the same way. I have been reading with interest anything I can find about seasonal and natural influences upon the body. This is something that has not been much studied. Certain people who work with people in institutional settings such as hospitals and nursing homes know that different behavior can be expected at different times, and it is not too great a stretch to credit day length or phases of the moon as influences. Any farmer who notices his livestock sees the same thing. We say the moon is wrong when the pigs run us over instead of sorting and loading easily.

The body's clock and calendar are powerful things. It seems right to be planting corn at eight in the evening in May, but sitting on the combine at the same time in October feels like you have already been working all night. Darkness makes a big difference. Maybe the sun regulates metabolism. Wouldn't that be a kick in the pants for all the biochemists who think they have us all figured out?

It has progressed to the point that I can no longer figure out snowbirds, that group of retired and semi-retired people who head for Florida and Arizona at the drop of a snowflake every autumn. Oh, I understand anyone who gets sick of winter. And even though I find the heat and humidity of July and August at least as hard to put up with as snow and cold, I know that the ice and wind of Minnesota's winter can be deadly, and that some folks are more at risk than others.

It is just that I don't understand what happens to the clock. If you go to warmer climates when the days start to get short, do you then start to sleep more even though the days aren't so short there? And when you come back, does your clock know that winter is over, or does it think it's still coming? How do they tell time in the South, where there is little variation in the days and the seasons are much less noticeable? Maybe they use watches and calendars.

If winter and fall are necessary so that we can rest after hard work, how do they do it in the South? Maybe they don't bother to work as hard any time of year, which is beginning to strike me as a very good idea. This idea, in case you haven't noticed, is all the rage in the farm press these days. Manage, we are told. Hire others to work, they say. They don't say if we can then expect to live longer or happier, but they do seem pretty sure we will be richer. If we manage, that is.

All the same, if I am going to do more loafing, I'd just as soon do it in a place where I know what time it is. I am certainly already confused enough without losing track of the seasons.

CONVERSATIONS WITH THE LAND

AUTUMN

September-coming-into-October is my favorite time of year. This is so because it is the end of the season that carries with it the promise that winter will wipe out all the mistakes we made, and we get to start over again next year. It is also the feeling of being involved in something much larger and more significant than ourselves. The small grains are in the bin, the barn is full of hay, the bedding has been made. The corn ripens in the shortening days and the fall winds.

We argue about the potatoes. Can we wait another few days? We know they will come out of the ground clean now. Will they if it rains next week? The freezer is full of beans and corn. We are enjoying the last of the watermelons and talking about how many years we may have to wait before we can grow melons that good again. The carrots are huge. Some of them will be frozen, some will just be covered up so that we can dig them some cold January day and feast our eyes on the bright orange color against the white snow and grey sky.

The two apple trees did not bear heavily this year, probably having decided to take the year off after producing like a small orchard last year. The apples are pretty good, though, in pies and various kinds of desserts. My neighbor presses a great apple juice from his trees. Even in an off year, the sheep enjoy the deadfall apples in addition to their regular task of cleaning up under the crabapple trees.

It is coming up on hog killing time. The spring pigs are going to market, such as it is. The sows are being readied for fall farrowing, while the sheep are being sorted by age and condition prior to their own late-fall breeding season. We are watching the hens and making judgments about them having to do with the number of eggs laid compared to our taste for chicken.

Meanwhile, there is a whole lot of closing up this season and preparing for the next going on around us that we have nothing to do with. The wild things see to what they need, and the perennial plants get ready to get over the winter. We have only to stay out of the way.

We have taken a little of our plentiful garden produce to a local nursing home this year and given it away. One old lady got one of our tomatoes, which she carried around and took jealous good care of for an entire day, showing it off and generally making it the center of every conversation. The next day, she divided it carefully into four parts so that her three table companions could also enjoy it for dinner. Some folks know what is worth having, and for some reason many of those folks are old.

Wendell Berry, Kentucky poet and farmer, notices wisdom from some really old people, quoting in one of his books some partial letters to the editor from a now-defunct Kentucky farm magazine published more than one hundred years ago. Here are some excerpts from the January 2, 1892 issue:

"No man…should spread his time and labor over so large an acreage as to fail in making a first class garden." I wonder whatever happened to that theory? Or listen to this:

"…It won't do as a rule to go into debt for…commercial fertilizers. You can more safely go into debt for good stable manure." Or:

"Nature never loses anything. It is only a fool man who squanders his substance and makes himself poor and everyone around him and the land that he lives on too."

This eloquent author, identified only as W.C. from Rural Neck, a place that no longer exists, finishes up his thought:

"When people learn to preserve the richness of the land that God has given them, and the rights to enjoy the fruits of their own labors, then will be the time when all will have corn in the crib, meat in the smokehouse and time to go to the election."

Leave aside the question of whether there is anyone anymore who wants to go to the election for whatever reason. We in this modern age who have been so very quick to give away any ability that we might have had to provide for ourselves, in favor of relying upon the experts and the big companies for the products the advertising geniuses tell us we must have, are not even going to understand the bit about enjoying the fruits of our own labors.

I know this. When I look at just the bounty provided by this small farm, its gardens and its simple barnyards and pastures, I find it incredible that anyone should be poor. And it is beyond belief that we have wound ourselves up in an economy so screwy that the bounty makes its producers poor.

CONVERSATIONS WITH THE LAND

HUNKERING

Winter came to western Minnesota before "official" winter this year. At this writing, we are living through our third blast, this one featuring high winds and fifty- to sixty-degree below wind chills. Official winter starts in about fifteen minutes, according to the radio.

I used to think that wind chills were something the weather services invented to terrify the uninformed and justify use of air time. If that is so, it hasn't worked. While the livestock take wind seriously and do their best to get out of it and try to live through it, we humans pretty much ignore it along with the rest of weather, and do as we please until we run into something or somebody.

Several weeks ago, we journeyed north to Moorhead to see our daughters, and it happened that the first wave of winter struck while we were driving. It hit up there first as often happens, and as we approached the city on the freeway we were warned by radio that freezing rain was spreading east from the city and that there were reports of cars in the ditch already.

The road surface was beginning to look suspicious to us anyway, so we slowed down to sixty or so, then fifty, and way down to forty as we got into glare ice.

To drive seventy or less on today's roads is to be an obstacle to most of the people driving. The cars came past in a steady stream. In ten miles we began to recognize those passing cars, now firmly planted in the ditch. One in particular, a sporty little red Pontiac, sat in the ditch, headlights looking back in the direction it had come, both of the young women in it talking on their cell phones.

Later that evening we saw that same pair of young ladies on the news, being interviewed from their car. (News crews never stay home, either.) The driver was not particularly embarrassed.

"It is SO slippery!" she exclaimed in wide-eyed wonder. "You have got to drive the speed limit!"

Right, I thought. Or maybe even less!

Something in us makes us want to run around in circles when the weather gets tough. We noticed when leaving Moorhead the next day that there are gate arms on the freeway entrances there. I can't imagine they keep many people off, but I suppose it is worth the effort. North Dakota talked a few winters back about charging stranded motorists for road service, which strikes me as a good idea.

Livestock, on the other hand, hunker down and wait. Sheep, if caught in the open, will drift with the wind until they get to a fence, which is a pretty good indication of the mental deficiency of sheep. Cattle will find the lee of something and stay there, not even venturing out to eat. And you ought to try chasing a pig out of a barn in this weather!

A good farm dog is a genius at coping. Of course she has been provided by nature with a good thick coat and a layer of fat in view of the approaching winter. She also is not tempted to use a breakdown-prone machine to get out beyond the limits of her ability to cope.

Our dog sleeps on top of the snow when the weather is calm. She has various places picked out depending upon precipitation, wind direction, and temperature. These are south porch, north porch, under the porch, in the dog house, under the tool bench, on the feed sacks, in the feed bin, in with the sows and in the hay shed, either under or on top of the bales. Those are the ones we know about.

I think that increasing age may be having its effect on me, foolish human that I am. When the wind howls, and as soon as I can be sure I have done what I can to keep the stock alive and healthy, I want to be close to the heat. Running to town and back looks a lot like too much effort. At home, the house fuel is up, the lights work or I can make them work, and the freezer is always full of food, as is as the pantry. There is a good supply of bread, sauerkraut and potatoes. There are books to read and thoughts to think.

So stay warm. Judging from the first part of this winter, we are in for it. People will soon be asking where all that global warming is at.

Today is the year's shortest day. As of tomorrow, the sun starts to win again, as the primitive cultures would say. Spring will come.

CONVERSATIONS WITH THE LAND

CHRISTMAS

Christmas has been linked in my mind with the daily winter evening trip around the lots and barns for a final check of livestock before turning in. It is as if all the nights I have done that, which is every night I have been home for nearly twenty-five years now, are rolled into one and tied up with Christmas in my memory. Forgotten are all the nights I found water frozen or a door or gate left or blown open, or when the walk resulted in a trip back to the house for more clothes and a quick cup of coffee to help fight off sleep for the next few hours while I dealt with emergencies from sickness to an immediate need for bedding. The scene that lives in my head looks something like this:

There is snow on the ground and it is cold. The stars shine with an intensity somehow unbelievable in the cold light of the next day. Even in this age of computer-enhanced illusion, nothing even comes close. If you want to see what I mean, you will have to get out of town late at night, and then stand there looking up until the cold seeps through all the layers you are wearing and you begin to know that you could not live through the night without protection. Then you will see.

I think this is a prairie thing. I don't believe the stars could be the same in the big woods. Here the night sky matches the vastness of the land, and beneath it human ambition pales to insignificance.

The Christmas story is a farmer story. I know that is why I have it linked in my mind with that nightly livestock patrol. Those barns and lots and pastures, with the breath rising in small columns of steam from all those mostly female animal noses, signify Christmas. Those animals are carrying the young that will be born in the coming spring, thus carrying out the ancient earth cycle of birth and death and rebirth.

Christmas is the personification in Jesus Christ of the cycle every farmer senses has been happening since the beginning of time. In another way, it is God speaking to us in the only way we are equipped to understand.

Or perhaps that should be "were equipped to understand." We have come a long way since several generations ago, when most of the earth's population made its way by one sort of farming or another. Whether our progress away from that state of affairs has been entirely desirable is an open question. Certainly more of us are now familiar with the virtual landscape of the internet or the hard plastic surroundings of the shopping mall than know intimately the ancient cycle of life and death. That fact can be understood as nothing other than a loss.

It is difficult for us farmers to handle the emotions and personal turmoil engendered in us by this jim-crack money machine we have become saddled with, and which now seems bent on destroying us. But handle it we must, because as surely as we are born, we need the awe and wonder that comes with a view of the night sky or the hearing of a well-known Christmas song or verse. Without it, we will become less than human.

What we need to do is celebrate. Even if you don't have livestock, get outside on Christmas night, or any other night. Take your kids with you. Take your grandchildren. If you don't have any, get someone else's. Show them that sky. Take pleasure in their wonder and share with them yours. Speculate with them about how far those stars are away and if they were always there. Ask them if they look the same from Jerusalem.

Then after a while you can tell them that this is the world they belong to, and that it is suffering from our lack of understanding. Tell them that they will need to try to do better at this than you have done, and that their health and wealth are linked to that big, big sky.

And if you are lucky, and if I am and if we all are, some of those children will begin to see that there is a life and a meaning and a big mystery to be part of and to try to figure out. Some of those children will begin to see the importance of what can't be bought but only marveled at, and then some of them will join us in living here under the stars that God spread across the prairie sky. If that happens, we will have saved some kids from the effects of some very bad modern philosophy, and started to spread Christmas into the rest of the year.

Merry Christmas from our house to yours! Take care.

CONVERSATIONS WITH THE LAND

DAKOTA

Visibility is again reduced on this ninth of January to the limits of the yard. I can just make out the outlines of the outbuildings from the house. We were warned, of course. The radio talked of falling temps and rising winds this morning, and several superintendents to the west sent the morning buses back home.

We set about doing the day's chores as well as trying to do a little of tomorrow's work ahead, knowing that what was hard today was apt to be impossible tomorrow. Livestock were fed a bit extra, and we renewed and added to bedding. By one o'clock we had given up on the planned hog delivery, parked the trailer, and went around closing doors and windows. At two, the snowblower tractor was plugged in, we were in the house, and the dog had crawled into the oats bin to wait it out.

This was all learned behavior, as we weren't born knowing it. Winter taught us that it was up to us; that if we want to live, we've got to think. The prairie suffers no fools.

Neither, evidently, does the governor of South Dakota. Most of the time I am pleased to be a Minnesotan. I think the state is ahead of the nation in some ways, and it generally seems that the advantages of being here outweigh the hassles. But there is what I can only call a "Dakota approach" that is breathtaking in its simplicity and effectiveness. Whenever it shows up, I am prone to jealousy.

Governor Janklow closed Interstate 29 this morning from border to border. He had already closed Interstate 90. The South Dakota governor went on to say that he had done so because conditions on the roads were not safe for the troopers, so they were not being sent out. He said that if people insisted upon going around the barricades and getting into trouble, the troopers would help out if they could. However, names would be taken and bills would be sent out after the storm. The bills, he thought, would probably run to about two thousand dollars.

That was it, straight and simple. If you insist on being a fool, the price of that foolishness is two thousand dollars. At least it is in South Dakota. Notice here the complete lack of "remedies." There is no pleading eloquently with people not to go out. No safe houses or warming shacks along the way. No cruising rescue teams or heroic snow removal efforts. Just a plain statement of intent. People being what they are, Janklow may have earned his year's pay with that announcement.

Just consider for a minute what a different democracy we would have if government at all levels took that Dakota attitude. Suppose the federal government told the plastics companies that they would be responsible for the nation's landfill problem because they created it. Or the chemical companies that they would have to figure out how to clean the groundwater if any of their product was found in the water.

This would be a world in which Northern States Power would not be allowed to generate any more nuclear power until they figured out how to disarm the spent fuel rods they have stored at Prairie Island. Major banks, savings and loan institutions, car companies, steel and oil conglomerates — bailout recipients all — would not be allowed to threaten the American people with moving their companies overseas until they had repaid the American taxpayer.

Outlaw companies would be told that if they couldn't afford to clean up the mess of their own making, they would have to go bankrupt. Can you imagine the American President telling Exxon to turn over the company checkbook to pay costs for the Gulf War? I can't. The President is not a Dakotan. The Dakotans are not that good at being Dakotan, either, as I am sure they blink if the fool has enough money, much as I might wish it were not so.

Oh, I know it's not as simple as all that. There is a cost to be paid by all of us if the powerful have to straighten up and fly right. Stuff in plastic will cost more, as will electricity and gasoline and crop chemicals and Chryslers. Yet it might be — should be — that our taxes would go down.

It would be tough, but not as tough as winter on the prairie. We survive winter because we have been selectively bred for it over three and four generations. A little plain dealing should be no challenge for us.

CONVERSATIONS WITH THE LAND

WINTER

The second Saturday of 2005, in the evening, is when we got the heads-up. My sister told us she had heard climatologist Mark Seeley on the radio talking about super-cold air slipping around on the polar ice cap. She said he wasn't sure it would come our way, but that if it did, it should arrive sometime late Thursday. Winter is something we always take seriously here in western Minnesota. By Monday morning they were talking about thirty-below on Thursday.

We weaned the December litters, which seems like a strange place to start with winter weather coming. However, several sows were easing off on their milking, so we needed to do something in the way of weaning a few anyhow, and it didn't seem like a good idea to wait out what might be a week or more of subzero weather before this particular group of sows got back to the boars.

By afternoon coffee time, we had the sows in the west end of our long farrowing barn and the pigs in the east, with the center section empty and pens ready to tear down. We had walked the fifteen sows from the building down to the end, with the exception of two late-farrowing sows and their litters that would need to go up to the other barn for the duration. There were two sows up there that had also farrowed later, so they stayed with their pigs. We also moved up one last sow that was yet to farrow, and then hauled the eight sows to be weaned down to join the others, as well as moving their pigs in with the others. We cleaned the east end of the farrowing barn, got in a big bale of straw, and grouped the pigs there with a portable torpedo heater.

After coffee, Josh ground feed for the newly weaned pigs. I hooked up a hayrack and went to get hay and barley bales up close for feeding cattle. When I got home I moved the springing heifers back from the corn silage to the hay and barley bale I had spread for them earlier, and opened the gate to let the spring heifers back at the silage. Then the newly weaned fall calves needed to be fed the evening ration. I made it to the house after dark, and Josh backed the feed mill into the shed somewhat after that.

Tuesday morning, they talked about forty-below on Thursday, and were beginning to speculate about wind chills. Tuesday is ordinarily a slow cattle day here since we do the bulk of the feeding on Mondays, Wednesdays and Fridays. We took the opportunity to straw the September pig litters in the hoop pretty heavily, and moved a few small, portable huts into the hoop with the two-month old calves so they could learn about getting in there away from the draft before they had to. Hoops get pretty cold in high winds, and the calves were apt to need a bit more shelter — especially since we had gotten this hoop late last fall from a neighbor, and hadn't gotten the east end constructed before frost stopped us.

In the afternoon we spent a few hours sorting the smaller pigs from the freshly weaned batch and getting them into separate quarters in the less-than-full barn. We took the smallest and weakest twenty percent or so, and moved them to empty farrowing pens in that barn, setting up heat lamps, water tank and feeders for them. I got more feed bales home just before the veterinarian showed up to work on a calf with a hernia. We had the calf in the barn in an empty farrowing pen, so we had lights to work by. This day ended two hours after dark.

Wednesday morning we had a detailed forecast to look at. Dropping temperatures Thursday. Twenty-five below Thursday night, with winds from the Northwest. Below-zero highs for Friday. Ditto for Friday night. Same again on Saturday. Our experience tells us that the warm-up predicted for Saturday night won't happen until about Tuesday, and that the northwest winds will be more westerly for us.

We decided to break the farrowing pens down in the center section of the farrowing barn, and stack them for the March farrowing so we could fill the building with hogs for heat. LeeAnn and I did this while Josh was hauling the weekly load of hogs into the butcher for the meat business. Then we moved in two big bales of barley and oats for several days of feed plus bedding, pushed the freshly weaned sows to the side of the west end, and got the March-due sows into the center part from where they had been living with the springing heifers. When Josh got home, we sorted the boars out from the March sows and put them with the newly weaned ones, closed the doors and called it a night.

Thursday, winter started to hit. We did the chores, which included putting out extra bales of bedding for the cattle at their windbreaks, plus an extra barley bale to each group for energy, along with extra hay. We were ready to push the spring heifers inside the sow barn if it got down to forty-below overnight and the wind came up. We never had to. By afternoon, we were getting in some serious house time.

As I write this on Monday morning, it looks like warm-up starts tomorrow. We have had no deaths or serious livestock problems for the duration, so we must have done it right. But looking back at this, I can see why we like grazing. We need the break from winter. Sixty days 'til green-up!

CONVERSATIONS WITH THE LAND

SEASONS

I stepped out on the deck at dawn one morning in March and walked across it to check the temperature. It hadn't frozen all night for the first time this spring, and the world smelled of wet cedar deck boards and mud. A pair of geese honked their way across the southern sky over pastures nearly half covered with standing water. Shreds of mist trailed up from the water as the morning fog began to develop.

There is a power that rises in the spring from the very earth itself. It is the foundation of a lot of human hope, some of it foolish to be sure, but much of it necessary. For me it is a constant source of amazement that it expresses itself in my own body and spirit; that while I stretch old, tightening muscles and settle with the pain in overused joints, I can feel the upwelling of the new strength from the earth in my own body. I need less sleep, generally dropping from eight or nine hours toward seven or even six. I am more observant, more alive to the world. But the primary element is the desire to push, to work at something, to make something happen. It is the expression of a primary life force, akin to joy. It is amazing to me that I feel it in my sixties.

Later in April the cows lean on the lot fence, gazing longingly at the first faint blush of green over the hills. Through the dead grass, last year's killdeer nests can be seen tucked into the smallest bump or groove in the earth. The western meadowlarks sound their melodious, multi-noted call from the tops of fence posts and small trees, and the vague stirrings of life begin to resolve themselves into work.

There is the water system to repair and start, fence wires to re-nail, machines to service for the coming crop work, a mountain of manure to haul, seeds to be purchased or cleaned. The days are longer now, and we race to get the necessary things done. Cattle are turned out to grass, and then the gestating sows. Pasture work gives way to oats seeding, then corn planting.

Soon we are doing weed control with drag and cultivator. The sun hangs in the sky for what seems forever each day as we cut and make hay in the heat, exhausted well before the end of the daylight but knowing full well that if we quit, it is we who will suffer the consequences. Farmer knowledge is direct and simple. Work needing to be done in April is not good enough done in May.

This is true if the delay is caused by weather, a machine breakdown, or by the farmer not getting out of bed early enough.

When the sun reaches its zenith in summer and seems to hold there for weeks, the power felt since the first stirrings of spring begins to wane, and is eventually replaced by something akin to fear. Soon fear will drive the work. It is a fear that the days will not be long enough to get the work done, that the frost will come before the crop is ready, that the harvest will not be brought in before the winter, that the animals will not have the time to put on adequate body reserves to make it through the winter, that the buildings will not hold out the prairie winds in January, that the house will not stay warm enough at thirty-below.

And so the animals are watched closely as they begin to lay on the weight from the grasses and forages that are full of the entire season's sun. Buildings are repaired, sometimes with patches on patches. Hay is stacked where it can be accessed through the coming snowdrifts. Feed is harvested or purchased ahead if need be. If there is extra time in this season of getting ready, it is often invested in getting ready for the next season. Seed is located, and new pasture subdivision fences built. Time is spent looking at the new gilts saved as replacements from the summer litters and speculating about how they will change and, it is hoped, improve the sow herd.

The last few days of August and the month of September are often a temporary reprieve from the primary agricultural drivers, as it seems that much of what will soon happen has already been put into place and must now be lived with. It is no accident that this is the time for county and state fairs.

Winter, when it comes with its frozen land and short days, is for regret and/or relaxation. If the prior seasons have gone tolerably well and if we are reasonably pleased with most of our own efforts, it can be a blessed relief from constant work — a time for vacation or just days off, for ice fishing, for breakfast in town, for visiting children, for sleeping in. It is also, inevitably I suppose, a time for planning, for there seems to be no helping the fact that farmers live in "next year country."

All of this cycle seems natural and acceptable to me, if occasionally harsh and unforgiving. After all, I have been a farmer since I wore diapers, and will be until I wear them again.

What I find increasingly difficult to understand as I age is the life with which I am surrounded. I do not understand "thank goodness it's Friday," or "pretty good for a Monday." I don't understand how anyone could think that anything precious on this earth could ever be protected by "insurance." I don't get how vast numbers of people who have never spent any time creating anything can participate in "recreation." I can't understand the meaning of the word "celebrity" or the idea that "education" should equal easy money. It amazes me that in the age of Bear Stearns and Bernie Madoff, folks still seem to accept the idea of money as a scorecard in measuring life. I don't understand the vileness I hear all around from the television and radio airwaves about people with a different opinion or belief.

Farmers are the tiniest minority in the United States, so we don't make much of a blip on the radar of the larger culture. All the things we could teach about connection to the universe that come through our close and yet conflicted relationship with nature and the land is unintelligible to folks on the outside. It will continue to be so to everyone who does not have the presence of mind to ask.

But I don't know how I would communicate to a non-farmer the feeling of power and drive that I get every spring. I am not sure it can be done.

CONVERSATIONS WITH THE LAND

CHARACTERS

CONVERSATIONS WITH THE LAND

KING

Topper was a German Shepherd bitch that belonged to Dad's good friend, Helmer. She had a batch of pups, the result of a whirlwind romance with a mostly Golden retriever, a mutt-about-town who lived most of the time with Cornelius (we called him "Corneil") just down the hill from our place and the other way from Helmer's.

Nobody ever had problems getting female animals bred in those days. It was not necessary for anyone to keep his own tomcat. We were a neighborhood. I remember a Holstein bull coming home from Helmer's, one step at a time behind a John Deere "B" in granny gear, with a rope attached to the ring in his nose. The bull enjoyed things a little more after he got here, as I remember.

The pups were born under the wreckage of an old "speedy" corn crib that had collapsed in its third season due to the effects of too much clay in the unwashed sand that went into the concrete. I begged Dad for one of them. We were between dogs, and I really wanted the next one to have those standup ears. Dad got around to mentioning it to Helmer…the day after the last one was spoken for. Typical for Dad, I thought. I was mad, and stayed that way for weeks. I wonder to this day if Dad ever noticed. He never said anything.

But things always change, even though a kid never believes it. The farmer who got the dog, the one I wanted so badly, died of a heart attack that fall. His widow decided that the thing to do was to move in to town, as she didn't think she could run the farm alone. Dad drove over one afternoon and brought the dog home. Dad locked him in the barn overnight to get him used to his new home, only to discover a major hole in the barn door on his way to milking the next morning. So there was another trip to the dog's former home to get him back. This time, the dog spent the night in the hay barn, and the next week tied with a chain to the granary. By the time the week was over, he was our dog.

His name was King. He had flop ears, courtesy of his sire, rather than the standup ears of his mother. And like his father, he thought of any unbred female dog in the neighborhood as a personal failing. Consequently he came home many mornings with ears bitten up, once with a two-inch tear next to the eyeball and numerous slashes and punctures in his shoulders. He didn't have the sense, my mother said, to stay home long enough to get healed up enough so that those ears wouldn't be covered with flies in the heat of summer. She had a grudge against him anyway. He was skinny as a rail, but still wouldn't eat the leftover potatoes she threw out unless she put a little gravy on them.

Tying the dog up to keep him at home is something that never occurred to us. The wonder is that no irate farmer ever shot him. Maybe it wasn't for lack of trying. King died of old age on a rug in the garage.

But while he was alive, what a dog for a kid to have! He could jump into the hay barn through the floor level door to the outside, a vertical jump of at least six feet. He clawed his way up the last foot and a half.

He loved to run. He kept pace with the car when we took one of our Sunday drives around the section. He ran with the tractor for days at a time, keeping the mice and hawks wary, and occasionally taking out after a jackrabbit. He trekked with me at six in the morning when I took the .22 rifle out after pocket gophers. As active as he was, he could wait quietly beside me downwind of the open tunnel.

I once saw him take on a badger. He lost, of course, as a dog generally does to a badger, but it didn't bother him any. He went home dripping blood, slept the rest of the day away on his rug, and was good as new in the morning. He learned to watch a gate while a tractor went through. That's as good as he ever got with the livestock.

For me the memory of King is a marker of the times. He was a mutt, something difficult to find today as we modern rural people keep much tighter control of our pets and desire the bragging rights a pedigree confers. The world he and I were born into doesn't exist anymore. Today I am surrounded by border collies, those bright young overachievers of the dog world. I love them for what they are and what they can do, which is little short of amazing when it comes to any kind of livestock.

But there is another dog around here, the result of a golden retriever's dalliance with the neighbor's border collie. She is a completely different dog from King, but Pepper is also my kind of dog, friendly and sort of dumb. She is hopeless with the livestock, thinks she can safely play tag with the tractor and skid loader, and barks at the blackbirds in the trees. She also greets me every morning, bounding over from her home on the old well cover and rolling over to get her belly scratched. She is just dog enough for an old kid.

CONVERSATIONS WITH THE LAND

LAUGHING DOG

Due to their habit of panting instead of sweating in the heat, dogs seem often to be laughing. So it has been, ever since the first canine crept in close to the fire of some Adam, that the dog has had to dodge the occasional well-aimed stick. Man does not take easily to being laughed at, and is easily provoked.

We are dog people in our family. Cats we tolerate because we know that they are an important part of the farm's food chain. We have a colony in the hay barn, and another bunch in the loft over the sow barn. We don't feed them; they do that for themselves. Come right down to it, as arrogant as cats are, they probably think of themselves as tolerating us.

Dogs, on the other hand, are constant companions, spirit lifters, service providers and, when they sit on their haunches and laugh at me, wonderful antidotes to pomposity and self-importance. This work they do nearly as well as my wife, and she might be thought to have signed on to the job, where the dogs just fell into it.

Pepper, the current dog in residence around here, is a weird combination of shepherd and hunter, which she expresses in the most useless ways. What she might be good for, in the range of things available for a dog to be good at, appears to puzzle even her. When she is around livestock and could be at least minimally useful, the Labrador part of her comes out, and she chases through the middle of the bunch. The lab is also in full play each and every night when she drags at least one longtime dead carcass onto the lawn.

When she is out running with the tractor, which she loves to do, the border collie part is apparently in control. Just last week she stopped in the middle of an old badger town in the alfalfa field, where a fox and six pups had just ducked underground. She sat down there, scratched her ear, and watched the tractor for a bit before getting up and plodding away. Later, while I was loudly cussing her for a fool from inside the tractor, she took off after a jackrabbit and tried to chase it into the next county. This farm produces lambs and some chickens, and could use a dog that would keep the local colony of foxes light on their feet. Jackrabbits are not much of a threat. Besides, it would take two of her to catch a jack. And if she ever did, the border collie in her wouldn't know what to do with it. With all her shortcomings, she is the first animal I speak to in the morning, and she never fails to walk with me on my nightly after-dark rounds. For that I will cut her some slack. And she laughs at me. That is worth something.

That is worth a lot! Listen…she runs with the tractor. I don't know if this particular dog is smart enough to make me understand, or if I am managing to get a little smarter myself as I age. The little tractor work that goes on around here, what with the big experiment in pasturing and the lack of patience for the road-farming we used to do, happens close enough to home that the dog can run with it at will. And I have noticed something.

When we go out, she runs gaily for at least three or four rounds. She is regularly distracted by flocks of birds that need chasing, as well as by the need to ram her head as far as it will go down a gopher hole. But all in all, she paces the tractor for those first rounds. By the time we get an hour or so into it, though, she is beginning to change her mind.

First she fails to make it all the way to the end, stopping instead about two-thirds of the way and whiling away the time until I return. After several turns at this, she sits down on a hill somewhere near the center of my endless back and forth trip, and watches me. After a few passes she might even lie down. The picture that stays with me is the dog sitting on her haunches, tongue lolling out, laughing at me!

A good dog can make a round or two in the field just for the heck of it, because she feels good and because the morning is cool and because she might run into something interesting. But after awhile, she cannot see the use of going somewhere that you are just going to come right back from. Eventually she goes home and lies under the deck.

These last years, I cannot get the laughing dog out of my mind. Do you suppose she listens to the markets?

CONVERSATIONS WITH THE LAND

DOG AS KID

Pity the poor mutt that happens to be in residence when the last kid leaves. The unfortunate canine is apt to be invited in, the first dog to cross the threshold in twenty or more years.

She had no idea she would ever be expected to be a kid substitute. There she sits on the rug peering into the kitchen, looking for all the world like an agricultural economist who, having stumbled into a diner full of farmers, is now expected to explain himself.

The warm air begins to take hold on the personal storehouse of barn smell she carries. Soon the atmosphere is filled with the olfactory reminder of every gate she watched, every sow she "helped" move, every tidbit in the corner of every barn she sampled in the past week. It is perhaps a measure of the heavy-duty changes life imposes on folks that the smell the farm dog brings into the house gets to be less noticeable when the house gets lonelier.

Life is ordinarily pretty good for the farm dog, the mutt who doesn't do much of anything very well except friendliness. You hope she barks when somebody comes up the driveway, and also that she is not over pestering the neighbors too much. The stock dog does the work. The ordinary farm mutt provides the love, affection and something to yell at.

She can occasionally and quite by accident be made to stand in the right place for a while when working livestock. Generally she will stay there until someone says something nice or encouraging to her. Then she comes over to be petted.

The dog can be counted on to find something dead every night and to drag it up close to the house to chew on. These are the same lips, by the way, that are right under your nose when you sit on the porch to put your shoes on in the morning. Whew!

Dog food from the store is nearly beside the point for the enterprising farm mutt. So are collars, heated doggy shacks and little sweaters. She is expected to provide her own.

She lives on the porch right outside the back door until it gets to be twenty below. Then she moves six feet over and crawls into a plastic dog box with straw inside. If a blizzard whoops it up, she may move in with the stock for a few hours.

In return she gets to run pretty much uncontrolled through the fields and pastures, whiling away hour after hour chasing birds and trying to dig out a badger. She chooses her meals, spends her day as she chooses, sleeps whenever it comes on her. It is the life of Riley. From the point of view of the farmer, she is a given. She is always there, like the machinery that needs fixing and the livestock needing to be fed. First thing in the morning, she is apt to be tripped over as she expresses her joy at seeing someone. She is not there whenever she could be useful or, if she is there, she is in the way. She trots right in front of the tractor wheels, tempting fate.

The kids think of her as company halfway down to the school bus in the morning. After they get licenses, she sits patiently while they get the old heap started, and then trots halfway down the driveway in their dust. She is always ready to listen when they need to tell someone about how stubborn and unreasonable dad is, or how difficult it is to get along with "that bunch" in school.

And then the last kid leaves. Suddenly, there the dog is, on the rug and making nice.

Oh, it's not that there are no advantages to being inside on occasion. There is the smell of cooking meat on a regular basis. She gets to eat some of it now, and sees how it has its advantages over her method, which is to wait until the sun and heat work their magic before she dines.

She gets petted more often. In fact, for the first time in her life, she gets some undivided attention. She is not entirely sure what to do with it. It may be that dogs are natural family counselors, with the advantage over the human variety in that they know how to keep their mouths shut and are therefore in less danger of really screwing things up.

But counselor or not, I think most dogs that start their lives outside would just as soon stay there. Inside is too confusing, and entirely too full of unrealistic expectations.

CONVERSATIONS WITH THE LAND

DOGS

Our farm has more or less become home to two medium-size black dogs, both female. There is Maggie the border collie and Pepper the perpetual adolescent. They spend most of their time getting in each other's way, but the collie actually does get some work done.

In fact, border collies must be the workaholics of the dog world. Border collies resemble what I imagine a junior executive with a shiny new Harvard MBA to be. The slavish attention the MBA pays to both the financial pages of the Wall Street Journal and the level of liquid in his boss's coffee cup roughly corresponds to the collie's absolute attention to the human in charge and the livestock to be manipulated.

Which livestock makes no difference. If no sheep are available, chickens will do, or even sparrows. They are always right at hand, and always at full attention. Mostly they don't know how to play. Many people find they have to keep them tied when they are not working just to keep them from going out freelancing with the stock. We have not yet found that necessary, though I can see what brings it on. They are like the MBA again in that they live to work, with little or no regard to what any self-respecting employee (or dog) would see as a private life.

I admit to being a border collie fancier. They are absolutely uncanny in their intelligence and ability. I was hooked by Lil, who came here five years ago or so. When she died early, I didn't think she could ever be replaced. Maggie comes close. It would be a stretch to say that they don't even need training, but not by much. You have most of it done if you can get them to come when you call. They circle by nature and will become accustomed to whether you want the stock brought to you or driven away. Sometimes they need a little help slowing down or circling far enough out to avoid rushing things.

Someone who knows said that border collies are very closely related to the wolf; that they were bred from the part of the pack that drives the prey toward the killers. This means a collie collecting livestock from a pasture is bringing them to the leader (wolf or human) who is expected to kill one, and then let her in on some of the chow. It must be a frustrating life for a dog! Sheep die often enough quite unexpectedly and of their own free will that perhaps the dog doesn't notice.

CHARACTERS

Then there is Pepper, a border collie-golden lab cross who is only about six or seven months old, which is perhaps why she plays for a living. We hoped she would have enough instinct to watch a gate. The jury is still out on that, but it's just as hard to get disgusted with her as it would be with a teenager who still thinks surfing is a life. Part of me hopes Pepper is right!

The two make an interesting spectacle when we get into the yards and pastures to move something, or even when just taking a walk. The border collie immediately circles the flock or herd and begins grouping them tightly, preparing to move them somewhere. Pepper plunges right into the middle, trying to find something to play with. The animals scatter. The border collie rounds them up; Pepper plunges in again. About here I yell at the pup, who looks at me and wags her tail, while Maggie the border collie comes up looking like the sorriest dog you ever saw.

This state of affairs is not beyond the border collie's abilities. A good collie can circle up a dozen sows in an outdoor lot and hold them in a tight group indefinitely. Anyone who knows hogs knows what a job that is! The difficulty is that two dogs at cross purposes are hard on the livestock. Since we like to occasionally make a profit with the stock, we don't let it go on for long.

Border collies don't bark while working. This is another wolf trait: they are soundless. I don't know if they howl. They will yelp when they hit the electric fence. So do I, for that matter. As a rule, any barking in the pasture is Pepper, and most barking by the house is Maggie. Pepper barks because it's fun, Maggie because she is high-strung. Probably neither one of them could be depended upon to bark a salesman off the yard.

If you admire a good dog, you haven't lived until you have seen a collie hold a sheep by looking at it. Lil, our first border collie, was what is known as "strong-eyed" — much more so than Maggie. She could hold a flighty sheep in place while we walked up to it. She went down flat on her belly, with muzzle along the ground, and glared up at the sheep. The poor critter was too confused to move. We could actually walk up to the sheep in an open area and grab it. This doesn't work with pigs, no matter how good the dog. Pigs have too much brains, I suppose. Dogs are one of the really nice perks that go with livestock farming.

CONVERSATIONS WITH THE LAND

TRUCK

One thing is that the oil gets changed a little more regularly now that the speedometer cable gave up the ghost and the needle lays quietly over against the 120 mark. He changes it now on the same schedule as the forty-year old chore tractor — every spring and fall, whether it needs it or not.

Other than that, the livestockman's pickup suffers a not so benign neglect. The difficulty is that the stockman expects the truck to be loyal and dependable, demanding little or no attention, something like his wife. But the pickup is just a machine, with none of the human capabilities for evening the score.

It started out as a flashy embodiment of someone's American dream, with a two-tone paint job, tires with white lettering and lots of chrome. It was lovingly washed every week, and various small gadgets and accessories like compasses and fuzzy dice were added.

But after a few years, the owner began to be attracted to certain new models. As soon as the payment book was under sufficient control, he traded "up," as he was pleased to think. After four months on the dealer's lot and a series of hard bargaining sessions involving the need of a heavier back bumper and a good, spring-loaded livestock trailer hitch, the machine fell into the hands of its final owner.

It hasn't seen a water hose or a can of wax since. About half the chrome is gone. What remains is loose either at the front or back end, depending mainly on where the rust has gotten the best hold. The tailgate has been used in a shed for the past three years to keep the salt and mineral blocks up off the damp concrete floor. There is no longer anything resembling a muffler.

The bed has developed a double pot belly due mainly to the way the sledge hammer, posts and various fencing tools have been thrown in over the years. It didn't help that the truck was enlisted to haul away the big concrete chunks left over when the barn-lot waterer had to be dug up and reinstalled a few years back, or that it regularly hauls home three tons of protein pellets in bags.

Every winter the bed becomes home to a pile of snow and twine strings from pasture feeding. These build up and accumulate all winter and spring, spending the last three months rotting at different rates depending upon rainfall and temperature. The composting ends when he pushes the twine out to make ready for the year's first fencing job.

CHARACTERS

Also found back there are a few ends of livestock marking crayons, a half dozen or so plastic syringes, a hog nose holder and an empty box of hog rings. Two hitch pins, seven pop cans and three styrofoam cups roll around. A few empty feed bags, a small watering trough and a half-full jug of iodine round out the inventory. No spare tire.

The right-hand outside mirror hangs limp against the side of the door, the result of the heifer tied to the back bumper that lunged over to get away from the vet. The glass shattered and fell out, so the black mirror background now stares blankly back at the driver whenever he looks that way. She left a good-size dent in the door, too. But it still opens, so no harm was done.

Inside the cab, you can tell what season it is by the smell. The cloth upholstery the original owner paid good money for has a big stain in the middle where a wet, half-dead newborn calf rode home at two o'clock one stormy March morning. The heater control is connected to nothing in particular and, rather than putting in a new cable, he has taken to turning the heat on and off under the hood each spring and fall. For fine tuning, he uses the window, the one on the driver's side that he patched a crank back on.

The dashboard has a crack in the plastic running from back to front, pointing to the stone chip in the windshield that has a crack spreading over to the right-hand side. These cracks serve as approximate division lines.

On the driver's side, the dash is full of tools — everything from a side cutters to a hog-ring pliers, to a fencing pliers and hammer, and even a crescent wrench and two good markers, one red and one orange.

The other side, which forms a sharper angle with the windshield, is stuffed full of receipts and bills. There are invoices — from the vet, from the elevator, from the parts supply house — a few offers to join record clubs, and one unopened envelope with the word **Giveaway** on the front.

The glove compartment contains every cab card and insurance voucher from the last ten years. On the passenger floorboard rests a pair of five-buckler overshoes with one buckle missing, along with a rolled-up pair of striped coveralls.

He has leaned on this truck to look at a healthy ewe nursing twins and to admire a really superior boar. It simply never occurred to him to turn around and admire the truck. His mind doesn't work that way.

CONVERSATIONS WITH THE LAND

STEEL WHEELS

I never burnt a hayrack. This is an omission and a regret; a gap where there should be a tooth, so to speak, as there were several racks in my history deserving of fire.

For example, there were two hayracks when I was a kid. My father's contribution to the neighborhood baling ring was part-ownership of the baler and two racks with our neighbor, Helmer. The racks were nailed together out of whatever lumber Dad had, and patched over with more of the same lumber as needed. You could sometimes walk from the front to the back without falling through the floor. Dad pointed out that the hay didn't fall through.

They were an embarrassment to me. The two of them, together with a like model owned by Helmer, made up our baling rig. The three racks, matched in dilapidation, rocked across the field — two of them on steel, one on rubber so old that an extra wheel and tire were needed in the morning to ensure a full day's work.

Dad preferred the steel-wheeled monsters, as they had no tires to go flat. These were old four-wheel running gears that would not carry the weight of the new, one-hundred bushel grain wagons, and so had been retired to the hayfields. When loaded they twisted, and the old lumber magnified every bump until from a side view the tiers of hay looked for all the world like the wind bellows on an accordion pumping out a polka.

"Jake, have you got the front tied down on this rack yet?" Helmer yelled while the hay fork holding ten bales left to climb toward the widow's peak atop the barn, where the carrier would trip and move it back into the storage area.

"If it stays down, then I do," Dad yelled back. It didn't, of course. The rack front ascended slowly, following the fork load of hay, while Helmer clambered down and I died a thousand deaths sitting on the tractor that was pulling the hay and fork into the barn by means of a pulley and rope.

I yearned for the end of the day, the end of the season, the end of something. Especially I hoped for the end of the old baler and the broken-down racks, and for Dad to take his checkbook down the road and hire the work done by the neighbor's son who was trying to start farming by doing custom

CHARACTERS

baling. He had a nearly new tractor, a new baler and four new racks he had built the previous winter.

Those racks were works of art. They were built on eight- and ten-ton wagons with new implement tires. Marvin had spread the wheels on the wagons, separating the front wheels from the rear by a considerable distance so that the long hayracks would be stable, and then he had painted them.

The hayracks were made on four-by-ten inch creosoted beams twenty-feet long, and crossed by eight-foot planks of treated lumber, brand new from the lumberyard. These were trimmed all around with light angle irons carefully angled and welded at the corners to protect the wood at the edges. The backs were tall and sturdy enough so that they moved with the rest of the apparatus, instead of a half-beat behind like Dad's. The monster racks would hold one hundred bales four high, or one-fifty if you stacked up.

Of course life only seems to always be the same. Helmer moved from his rented farm, buying our half of the baler and taking it with him well before I died of embarrassment. Nearly forty years later, one of the steel-wheeled wagons quietly rusts and rots in the nearby grove. Helmer and his beloved Rose, as well as my Dad, are gone from us. And today I pulled the nose off my grandson's face and showed it to him squeezed between my fingers. He reached up quickly to check his nose, just as I had done when Helmer pulled that trick on me so many years ago.

I eventually got to work on those wonder hayracks, for I was growing up and Marvin paid one cent a bale to each of the two teenagers who alternated stacking loads and driving tractor. It was big money. Because Marvin's machine had good lights and because he was ambitious and because it was a new day in farming, sometimes we would come in to the yard and help the farmer stack the last of the bales, go in for supper and then, if there had been no dew and there was a breeze up, go back out and bale everything full again, finishing at midnight.

Hayracks, those first embarrassing ones as well as those built by my neighbor, are a kind of pivot point for me. You see, I never looked back from the time I first got that work on those new hayracks, all the way through my growing up and the first twenty years of my farming after the university interlude. Newer was better, and I was a technologist. I still am, for all that. But a few years ago, while I was still running I know not where, it was as if I stuck a pitchfork in the ground and held on while that pivot threw me around in a hard circle. And while I was circling, one of the things we did was build racks.

37

"We," I say. Josh, my son, built the backs out of steel tubing with neighbor Marvin's wire-feed welder while he worked over there after high school. And my sister, who is a pretty good carpenter, built the wagon floors, I think, to work off some of the rage she felt over being downsized by her employer. I helped her with the nailing the last day of the year after Dad died. I felt him at my elbow, swinging that hammer as he built steer troughs in the winter, feeding fence for the cattle in the evenings, and a bench for his daughters to sit on at the table. A whole host of us worked on those racks for a whole host of reasons.

And what did we build? We built a new direction. The hayracks, while virtual copies of those I so admired forty years ago, have a whole different meaning. Forty years ago they were a custom worker's tool. Now, they stand for a resolve to move a row-crop farm toward sanity. In a world of specialized machines, they are low-tech and general purpose. They were built on wagon gears no longer needed for the grain boxes because the farm is not planted exclusively in corn and soybeans anymore, and because the grain harvest is now a cooperation with a neighbor who provides plenty of hauling equipment.

In a world where everything is bought, they were cooperatively created. They are an acknowledgment that everything worth having cannot be had by chasing after it with money.

And they are certainly the closing of a circle, a nod toward those to whom much is owed, and a shouldering of the responsibility to give the future something worthwhile in exchange.

CHARACTERS

MORT

You need to understand that Mort was a lifer. He was in the military for awhile, earlier on in his time. He must have been a lifer there, too.

Mort wasted most of his life waiting for it to begin. He waited for retirement and he waited for winter. He waited for Friday and he waited for the end of the day. Sometimes he even waited for noon.

Mort was the first lifer I ever knew; I just never saw it among the adults who raised me. But when I arrived for work that first summer with the construction crew, Mort was a revelation. He had a paunch that would have comfortably housed a couple of good-sized watermelons. And because of the way he could lean on a shovel, the wooden handle of the tool neatly bisected that belly, folding it most of the way around until the handle almost disappeared between the place where his knee leaned against it to where it appeared full-size again, just below his hands. The hands he kept clasped one above the other, with the top one featuring the forefinger lying atop the handle. I suppose that whole posture minimized the effort it took to remain upright.

Neither hand was necessary to operate the Pall Mall that was always in his mouth. Mort did that with his lips and lungs.

The miracle was that he had the reputation of having a sort of sixth sense. Mort knew where the boss was. Many was the time when the only rock I saw the man move all day, he moved while the boss was driving by.

Economy of motion was something Mort understood. And he was good at it.

One day in the middle of the week we woke to the sound of steady rain on the roof. The hope of a whole day off work got us kids on the crew wondering how far the nearest city was and who would drive. From his corner, Mort growled at us to shut up, and rolled over for a few more hours sleep. By the time we had breakfast in the cafe, finished admiring and criticizing each other's cars, and gone over for the hundredth time the short list of high school girls who had put in an appearance so far around the construction site, the rain had backed off. The sun peeked out. With the temperature and the humidity chasing each other into the nineties, the word came around that we would work the afternoon. We were to flush Shorty out of the saloon where he had been steadily throwing down boilermakers since ten o'clock.

Think of Shorty as a Mort in training. He was a couple years older than us, but he wasn't headed for any college. He favored ducktail haircuts and white tee-shirts with the Camels rolled up in the sleeve. We got him out to the job, put him behind the wheel of the truck and propped him reasonably close to upright. We figured he could operate the clutch and super-low gear even in his condition. We started to pick rocks. We were kids; we didn't have much judgement.

We didn't have much in the way of mercy, either. By the time we had been out there an hour, the temperature was climbing toward one hundred, and soft moaning noises began to be heard from the cab. Shorty was coming out of it, complete with a rip-roaring hangover.

One of our number, a blond giant of six-foot-five, picked up a rock about the size of a cantaloupe. Windmilling the twenty-pound chunk over his head in the best Camilo Pascual fashion, he strode into the pitch and slammed the rock into the wooden upright built into the stake bed right behind the cab. The posts gave, and the entire backstop smacked into the back of the F-6's cab with enough force to cause dust to fly out all around the rear window seal. The cab echoed like a tin can.

For a moment there was dead silence. Then the driver's door swung open, and Shorty rolled out over the running board and into the ditch holding his head and whimpering.

As Shorty lay in the ditch getting rid of the last of those boiler makers, Mort lit another cigarette. "Kids," he said, as he scratched the match to life with his thumbnail. He shook his head. "You kids!" Two of us walked into the ditch, grabbed Shorty, and put him back into the truck.

CHARACTERS
ELLIS

My friend Ellis was a farmer to the end. He died several years ago by drowning in a corn bin at age eighty-three. He had been getting increasingly frail, even skinnier than usual as he spent his last years of life on his farm, and Sundays with his wife in the nursing home. Toward the last, he took to visiting the Veterans Hospital every so often for an IV, getting himself, as he put it, "pumped up a little again." His teeth then fit better for awhile, and he didn't click and clack so much as he talked.

Because he was a farmer of the older sort, he understood a lot that many of my other neighbors did not. He caught on to where I was headed, often without being told. I remember the first fall I set bales out in rows in the pasture for feeding the cattle there during the winter. The pickups were causing a traffic jam out on the county blacktop in front of the field, as various neighbors and community know-it-all types tried to figure out what I was up to this time.

Ellis came bombing up the driveway in the old minivan that served as his pickup, and told me with a twinkle in his eye that he saw I was up to getting the manure on the pasture the easy way. He sometimes knew I had a new litter of pigs in the pasture before I did. The idea that we were two generations on the same farm, as were he and his son, thrilled him to no end, and he would do whatever he could to encourage it along. And though he was a crop farmer, his feet were in his own history, and his mind was not foreclosed to the possibilities inherent in our efforts to do a different thing.

Ellis was a strong Democrat, a true believer in the capacity of politics to move people. He was a prairie populist, as fierce in his beliefs as that sort can be, but with a childlike belief in the capability of his neighbors to see things his way. He was hard on the wealthy and powerful in his talk, but gentle and forgiving with everyone else.

He lent me the use of his loader tractor once. I don't remember why I needed it. It was a pretty casual affair, a lightweight New Idea loader on a 706 gas model IH. As he put it, the tractor was "temperamental." Ellis was of the generation that made the switch to tractor power and still thought of them as horses with wheels, full of personality and stubbornness. You know the ones. Very few of them got to the cemetery with all their fingers and hands. I grew up thinking it merely normal to go to church with a certain number of men sporting hooks where hands should have been.

"Choke her hard," he told me. "Then when she starts, leave the choke on full for awhile 'til she gags and sneezes. Then shut it down to half and hold it there until she misfires. Then push it in. Don't worry about the smoke. She gets over that, too, after awhile."

And it was Ellis who gave me my first idea of the height of the hill we were trying to climb with the marketing, telling me that he had tried selling eggs when he was younger, but that the people wanted them not for what they paid in the store, but for what he would have gotten for them at the produce in town.

Ellis was raised in a church that took itself and its beliefs pretty seriously. There was a church a few miles away with its own cemetery where he and a dozen other farm families went. After service, while Sunday School was going on, they would go to coffee at one or another of the nearby farms. After an hour of farmer talk and checking out the cows and the sows and the corn, they would head home by way of the church to pick up the kids. Next Sunday it would be a different farm.

When they finally closed the church thirty-five years ago, the congregation joined the one of the same denomination in town. After service, Ellis told me, they met at the McDonald's. Not the same, he said sadly, with a shrug of his shoulders. And the preacher preaches politics, he said.

In the last year of his life, Ellis left that church in town and joined his son at the last remaining country church in the neighborhood, one of a considerably more liberal frame of mind. It made no difference to him. He was happy there. The place is full of farmers, he said with a quiet smile. For Ellis, there was still the possibility of redemption, even the redemption of a dreamed-of life on a farm in an almost ruined farming community in his last year on earth.

Ellis thought that farmers should want to live mostly on the farm instead of at the lake or the casino, and he could not, for the life of him, figure out why so many who surrounded him in his later life did not. He loved it all, from the rain that wouldn't come or wouldn't stop when it did, to the trees that scratched his tractor, to the crop that amazed him one year and disappointed him the next. He loved the sweat and the tears and the loss of it, because he knew that without those drawbacks he didn't get the new life, the

CHARACTERS

new shot at it the next year, the dew on the grass, the birds in the trees, the possibility of rest and the comfort of neighbors.

He was, in this way, a complement to my own father, who told me when I was sixteen that of course everyone wanted to farm, but that unfortunately there just wouldn't be room for everyone, so some of them would just have to be satisfied with doing something else.

Ellis was buried in the cemetery next to where the country church stood, the one he attended for so many years among friends, relatives and ancestors. There are a few cottonwoods nearby for music when the wind blows, and he can keep an eye on the corn.

CONVERSATIONS WITH THE LAND

TRUCKERS

From south of the barn there came a yelp of surprise, followed by a string of the vilest profanity I had so far heard. Dennis looked at Dad and said, "Joe found the fence!" Mom said, "Oh, I wish he wouldn't do that." Then to Dad, reproachfully, "Jake, didn't you pull the plug on the fence?"

Joe was the first livestock trucker I ever knew, and my memory goes back to a wonderful trip sitting wedged between him and my father in a 1955 IH binder for 125 miles hauling cattle into the stockyards in South St. Paul, overnight at the Drover's Club, and then back in the morning. In my mind I can still see those city lights spread out as we topped the river bluffs in the evening. Joe popped it into third, shifting the axle at the same time, and we roared down toward Concord Street and the yards, the load of cattle pushing us toward the river. Joe gobbled and stuttered when he talked, and the loose skin that was kind of suspended from his neck would wobble when he moved his head, adding to the turkey illusion.

The more excited he was, the worse he gobbled. Joe was near to unintelligible that afternoon when he grabbed the hot wire to swing his leg over. But the profanity still stuck out loud and clear. Dennis, his nephew (great nephew, actually), had sent him down to get behind the last several cattle and bring them up to the shed for sorting. Part of the reason for this was that the old boy was pretty excitable when it came to livestock, and he fancied himself still a young man when it came to wrestling a twelve-hundred pound steer onto a truck. Dennis sent him down there to keep him from getting hurt.

But Joe found the fence. He had a reputation for finding fences the hard way. It is a wonder to me now that he could be trusted to drive the load into the city, but not to load them. Probably Dennis had been run over once too often by an animal that Joe had spooked.

Livestock truckers are not a tightly knit breed. They are all over the spectrum politically. Some swear continuously; some are pretty religious. Some are pretty casual on the time; a few are prompt. A few have a clue about the principles of handling livestock; most do not. A few are natty dressers, favoring cowboy boots and fancy shirts. These tend to show up in shiny black "dually" pickups pulling aluminum trailers. Most, though, have given over their fashion sense to the realities of their work. The most memorable are entertaining — to be counted upon to grab a hot wire — or for a little clowning or telling a story not heard before.

Tom tells stories. "I've got a niece, she thinks she's a vegetarian," he told me through his bushy mustache. "So there we are, having a good time all of us out to eat together after the baptism. Must have been twenty-five, thirty of us in the restaurant. All of a sudden she starts up whining that there is nothing on the menu she can eat," Tom related.

"So I took the menu away from her and spent a little time looking it over. Then I handed it back to her, pointing at the chicken section. 'There you go,' I told her. 'You can eat chicken. That ain't meat.'" Tom got quiet and thoughtful for a minute, remembering the look on her face. Then he burst out joyfully, "She thinks I'm a hick!"

For the third time, Josh and I brought the last of the butcher hogs to be loaded up into the corner. Twice before, the big black barrow had turned and snowplowed his way through his mates, me, Josh and the gate we were holding, racing back to the farthest reaches of the pen and followed by the rest of the pigs. This time we were determined to hold him. Sure enough, the black barrow tried it again. This time he wasn't able to get his nose down to go under, so he came back toward us, crawling on top of the others that had resigned themselves to loading. Josh laced his fingers together, got them under the pig's jaw and lifted, throwing the porker on his back, with his nose in the trailer. As he scrambled to turn himself right side up, we locked hands under his tail and pushed. Luckily enough, he still had his front legs folded underneath, and we skidded him on. Tom dropped the trailer gate just inches ahead of the pig's nose as he turned to come on his way back out.

"Republican pig," said Tom. I was interested. "Yeah?" I said, as I stood there using my hands to squeegee the manure off the front of my shirt.

"Only knows one way, and it's the wrong way," replied Tom. That was worth a chuckle in spite of the smelly shirt. "I get a kick out of that routine," he continued. "I tell it every so often, and if the guy laughs, he's probably pretty much a Democrat. The Republican guys don't like to hear it!" Then a pause while he thought, probably reminding himself that his little trucking company was, after all, a business.

"Doesn't matter though," he said. "I'm pretty nearly the only trucker around. If they want something moved, they've still got to call me."

And we have had a succession of cowboy types, all of which blend together in my mind. I will call them all Randy, since Randy was the one of that type who lasted the longest. The electric prod was his tool of choice, and he remained convinced that if the cattle hit the chute running, they would all end up on the truck, no problem and no questions asked. Any woman or kid unlucky enough to be holding up a gate at the weak point in the system would get run over when a Randy-generated stampede hit the wall. As the miserable individual got up, covered with mud and manure, Randy's comment would be along the lines of: "Aw, you let 'em through. Now we gotta chase 'em."

And as he pulled the rubber boots off over his cowboy boot heels, he would always add, "Well, it coulda been worse."

It was enough to make me imagine what catastrophes he had caused on other farms. We eventually convinced Randy that the end gate of his truck needed his close attendance and supervision at all times, and that it was more important he do the driving and leave the loading to us. It may really be that there is a place for everyone in the world, and that Randy's place was holding on to the rope that controlled the drop gate. After all, he was willing to do all that driving. And you know, at the time he was pretty nearly the only trucker around.

CHARACTERS

GEO

For certain jobs, the neighbors and a few relatives were around a lot. We hayed together and made silage that way as well. In my memory, the grain harvest and the corn picking were already individual farm operations, but silage remained a group effort for some time.

My dad owned the silage outfit, which consisted of our farm's main tractor, a "70" John Deere two-cylinder that was hitched to a single-row Allis Chalmers silage cutter, to which was hitched one of two lumberyard-built silage dump boxes — the green one or the orange one. He also had an Allis silage blower for use at the silo, along with about fifty feet of silage pipe.

This group of machines and about as many men went around each fall to each of the group's farms to fill silos with chopped corn. Unlike haying, silage harvest generally started in September after school was in session — a fact I hated because I was sometimes more farmer than scholar.

Remarkable in my memory is that this group took the time, after everyone's silo was filled, to come back around the loop a week or so later to refill each silo to the top after the original silage had shrunken a bit. The refilling took perhaps two or three loads of silage, but it also meant that everything would have to be moved down the road to the next farm again. The pipes would have to be brought down, taken apart and loaded on the hayrack for transport. At the next site, the pipes would be laid out again in the cow yard, where six or seven of them were assembled, four bolts to the joint. The bolts came out of a coffee can that surely had been rained into at least once during the season, meaning they were beginning to rust and would need the application of several wrenches and considerable energy to get them all the way tight.

Then someone would have backed the blower up to the silo, gotten the tractor lined up so the drive belt would stay on the blower, staked the blower down, and blocked the tractor wheels to help out the brakes. When the pipes were assembled and tight, someone needed to shinny up the silo on the outside ladder carrying a wooden and cast-iron pulley with the hay rope strung through it. This would be fastened at the top, with one end tied to the set of pipes on the ground. Two or three men on the other end would pull the pipe assembly up and into place. The man on the silo would have to adjust direction on the pipes so that the gooseneck on the end would indeed direct the silage into the silo. By this time, several loads of fresh silage would be waiting in the yard. After the silo was filled, the entire process went into reverse so that another farm could be refilled in the afternoon.

This was done because it needed to be. If it was not topped off, about ten percent of the silo's capacity for the season would be wasted. Labor wasn't thought of as a cost.

George was our high man. We called him Geo. His wife called him Geo, or sometimes "the mister." Geo was short, perhaps five-foot-five, and bald as a cue ball. He had no thumbs, both having been sacrificed in the effort to learn to use the new farm machines, with their belts and power shafts. Geo was very adept with what was left of his hands, having taught himself to use the fore and index fingers of either hand together for such things as starting nuts on bolts, holding nails to start, operating wrenches and castrating boar pigs.

News of the latest "inconvenient" birth in the neighborhood always called forth the same joke. "Has the baby got thumbs?" Geo would ask. "Because if it does, then it ain't mine!"

Geo was thumbless, but ambidextrous. And he did not fear heights. He would stand on the top rim of a forty-foot silo with only the drawn-up pipes to hold to for balance, working them into position to be tied off for the job. The crew on the ground held their breath, hoping for the best. Geo was comfortable up there, yelling down when he needed a little perspective on position, when we hadn't put on enough pipe, or maybe that he thought it was high time we got his tractor into position so we could get into the house on time for dinner so as not to upset the missus. He giggled continuously as he talked, maybe a habit picked up as a tough little boy learning to live in a world where everyone was bigger.

On the blower, we used Geo's old WC Chalmers with hand brakes. The "Allamus Chunkers," as Oscar called it. It gave a good, steady power for the belt, unlike the two-cylinder tractors that tended to throw the belt when the load got heavy, thus plugging the pipes and causing an hour or more shutdown to clear them. Geo brought it on the first day of the run with a nearly empty gas tank, and drove it home at the end with the tank full. It was one of his economies.

Another was that when he was done using the boar for his three or four sows, Geo would simply quit feeding it, turning it loose to wander around.

CHARACTERS

The boar spent this free time around the old drainage ditch and foraging the neighbor's cornfields, as well as sneaking past the dog to pick up a little spilled feed on a strange yard. If necessary, the pig would spill it himself. The wonder was that this boar generally showed up again around the time its services were required. Once in a while, it was early.

"Geo's boar showed up again, and was he ever surprised," reported Helmer. Geo probably thought somebody shot the pig. Shooting the pig was indeed thought about and talked about, but never done. Geo had a value to the group that was not defined by the price of a little pig damage, even by these shrewd and often hard men.

When the sheriff sold him out, Geo hid his John Deere "A," the plow tractor, behind the tall cottonwoods and honeysuckle down on the drainage ditch. Every neighbor knew where it was, but the sheriff didn't. Geo never came back for it. He was in town now, janitoring in the school. At the ripe-old age of sixteen I became the high man on the silo. The tractor sat on that ditch deteriorating for twenty years until a renter gave it to the junk man.

WIVES, HUSBANDS, CHILDREN

WIVES, HUSBANDS, CHILDREN

DEATH & REBIRTH

On Sept. 8, 2001, my mother, who had earlier suffered a major heart attack and stroke, broke her hip. On September 11, three passenger jets hit their targets in New York and Washington, killing thousands. I will never forget those days.

As Mom struggled through a morphine haze trying to figure out how to continue living on this earth, others had plotted, and still others were beginning to plot, the destruction of life. The hospital reverberated to the televised sound and fury surrounding the making of an American posse. I turned off countless television sets in those few days. I was determined that Mom should have the chance to leave this life without being terrified over the fate of her grandchildren.

We made and followed a "no television, no radio" rule in our home for the week following the eleventh. But there was no escaping the news. And so it happened that I learned how out of place I was in my own country at the same time I was faced with needing to learn about being the oldest generation in my own family. Mom gave up her fight and left us October 6, the same day her granddaughter was married.

As horrible as they were, it wasn't just the jets and the ensuing firestorm that bothered me. That was an act of war. But what we were coached to do next felt like going through Alice's looking glass and sitting down to tea with the Mad Hatter. As the buildings were searched and survivors found and the story began to be pieced together, we heard that the proposed war of retaliation would not require much in terms of sacrifice for Americans, that it would be more of an inconvenience. We were encouraged to travel, especially by air, and to plan vacations in faraway places. We were treated to the spectacle of tragedy heroes crooning "God Bless America" at the New York Stock Exchange. We learned that at all costs, we needed to keep up our spending. This would support the economy, we were told, and show those terrorists that they had little or no impact.

I thought of Mom's generation and what they had given up for the country by going off to World War II and, for those left at home, the sacrifices made to support the war effort. The dissonance between what I was being told should be our response to the terrorist attack, and what I had been told of life during WWII, was painful to me. Is this what our two-hundred year experiment in self government comes down to? Is it our contribution to peace and justice in the twenty-first century to turn ourselves into a police state so that we can *party on*?

A few days after my mother's death, my sister eulogized her with these simple, elegant words:

"Every two weeks or so, Mom and Dad would take their surplus eggs to the produce in town. Mom would do her grocery shopping afterwards from a list she had made, which always started with the 'must haves' on top, ranging down to the 'nice to haves' at the bottom. When the egg money ran out, the rest of the list stayed in the store. When she got home, she carried two weeks worth of provisions for a farm family of six hungry people into the house in two bags, one under each arm. And we never lacked a thing."

It is true; we lived like kings. My parents knew about provisioning. Mom knew about potatoes — when and how to plant, harvest and store. She cooked and baked with what she had on hand. She could kill a few chickens each week; she knew which ones weren't laying. She kept a garden. Dad milked cows and could slaughter a hog. They canned, they salted, they froze. And when they ran across something they couldn't provide and didn't have the money to buy, they quietly and courageously guessed they could do without.

If President Bush is looking for a way to point out America's indomitable spirit to the world, he could do worse than to point at my mother's life. Think of the safety and security to be had by following a few principles that appear to have guided her life:

• Don't buy what you can provide for yourself.
• Take care of your own.
• Don't be afraid of a little work.
• Insofar as possible, live in peace with your neighbors.

That kind of living looks pretty good right now as we face the dangers of adulteration of a food supply that we have so stupidly allowed to be centralized. Or the interruption of a central electrical power source upon which we absolutely depend. Or as we think about the asinine way in which we have allowed petroleum to become necessary for everything about our lives. We have given control of our necessities over to people who do not have our interests at heart, and all our arrogant pretense about living high and doing what we wish will not ensure our safety or our well being.

Mom's life, and now her death, have given me all the indication I need that the direction we have tried to take our farm this last decade — toward real wealth, stable family and community, and kindly use of the earth — is the right one. Best of all, it is a direction available to any who wish to learn and who are tired of waiting upon the pronouncements of opinion makers.

CONVERSATIONS WITH THE LAND

KITE

Few things satisfy like the sight of a kite sailing over a September pasture. The brightly unnatural reds, blues and greens of the nylon material contrast with the dusty, late-season green of the grass and alfalfa and the quiet blue sky. The effect of those colors and the vertical aspect of the string leading up to the kite is that of an exclamation point, a visual hope that humanity can belong in nature.

I hadn't flown a kite in forty years, and took some satisfaction in the fact that I could still run fast enough to get it off the ground. Forty years ago, kites were of paper and light rags, and there was more spring in my step. But this nylon and string affair was just the ticket for my grandsons, who were delighted with every zag and bounce the kite made in the undependable breeze.

We used to fly kites in the spring, choosing a day not windy enough to tear the paper, but sufficiently so to keep the thing up there for awhile. September flying was new to me, and it may be a benefit of some of the choices we have made around here.

This year I gave up the combine, figuring that twenty-five years was long enough to cope with that thing. The row crops that go with it are greatly reduced as well, opening some space for important things like kites (and grandsons). Not only did we have room to fly a kite but, since I wasn't sitting on a machine or up to my elbows in grease trying to make it run, I had the proper attitude.

The breeze strengthened and steadied after awhile, and it became apparent at least to the boy who just started kindergarten — and who is therefore terribly worldly — that the kite had enough lift to get him off the ground. He took a little convincing, for he was judging by the pull the gadget gave to the string, which was by this time considerable.

I told him to hand the string over to his little brother, who wanted a chance.

"Don't do it, grandpa. He'll just let it go."

"No," I said. "He'll do all right."

He let it go of course, giggling all the time the kite took to reach the ground. His brother ran after the string, which stayed just out of his reach, yelling back his opinions of little brothers in general and this one in particular.

After awhile we gave younger brother another chance, and he let it go again. Again it came down, proving, I guess, that kites can only fly free like a bird if you keep control of them.

It must be about flying, this fascination with kites among some of us. There are those who think our species pretty miserable because we are not able to fly. History is full of stories about people trying to remedy this, from the Greek myth about Icarius flying too close to the sun and melting the wax holding his feathers in place, to any number of instances of folks under the influence of one spirit or another throwing themselves from high places.

We have "solved" that problem with a machine that I for one do not particularly trust. My fear of airplanes, which I have demonstrated to my own satisfaction several times in my life, must relate to my general distrust of machines. If a machine is a gadget constantly on its way to breakdown, it seems merely good sense not to ride one several hundred or thousand feet into the air. Shinnying up a silo on an outside ladder doesn't give me too much of a problem, so it isn't a straightforward fear of heights.

We tried that kite in the middle of the pasture and over toward the cornfield, finding the breeze was less predictable next to the tall corn. We made it move by jiggling on the line, and then spent considerable time lying on our backs and watching the clouds move past the kite, seeming to put the universe into a motion beyond our control.

Find a field. Buy a kite. Or make one. That must still be possible. Then fly it. Get some grandchildren to go with you. If you don't have any of your own, borrow some. It is not possible to do a kite justice unless you have children's eyes available to see it through. Their wide-eyed delight and wonder will make you remember when the world was new for you that way.

This will help you be yourself in the world in a way that the shopping mall never will. There really are some things you can't buy.

CONVERSATIONS WITH THE LAND

BULLS

Linda Hasselstrom, that careful observer of ranch life on the northern Plains, writes about a careless coyote walking into the midst of a small herd of brood cows with calves. The cows bellowed in outrage, and their mates from all ends of the pasture came galloping to the rescue. In Hasselstrom's words, they "crashed and collided, pawed the ground, slung snot in the air... Two of the bulls squared off, and slammed their heads together. Confused about the nature of the threat, they fell back on masculine custom: if in doubt, fight."

Anyone who wants to loosen up a room at a sustainable agriculture or grazing convention can tell a story like that, with instant results. I often relate Alan Savory's observation that in his native Zimbabwe, when the men got called to war and left the farms in the care of the women, the farm businesses suddenly got profitable. Of course the reason anecdotes like this are popular is that grazing and sustainable agriculture include women. Conventional or industrial agriculture suffers, or connives at, their absence.

Perhaps this is a natural outcome of the "industrial" in the name. Men relate better to tractors and trucks than to pregnant cows. The overpowering of the "culture" part of agriculture has been in progress for most of a century now, and we are left with crops that grow alone in the fields, windbreak trees that are cut down because they might scratch the paint on the tractor, sows surrounded by enough metal to build another barn, and separation and compartmentalization everywhere. The landscape of rural America is a snapshot of the interior of a modern male brain.

But the facts still are that tractors and farrowing crates depreciate, while animals, especially brood animals, not only appreciate in value, but reproduce. Much has been made of the fact that we need the children if agriculture is to continue. Too true; but how do we keep the children without the women? Biologically, of course that is impossible — at least so far. But socially it is as well. If Mom is working in town to support Dad's farming habit, who coaches the kids to see the beauty in the bred sow, and the cow with calf at side? Or if the kids see that the joy of farming is bound up only in what can be painted and driven, why wouldn't they choose to work in town, where some company might actually provide them with health insurance?

WIVES, HUSBANDS, CHILDREN

This emphasis on all things male is a great issue that runs through all of society. It shows up in our nation's relationship to the rest of the world, in how we do our business and in how our technology is invented, grows and is used. In how our cities are laid out, and in the expectations we have or do not have of our corporate structure. It is there in our crime — in what gets punished and what does not, in our expectations of our relationships, and in our hopes for the future. Change in this area would not come with a woman in the White House and some idiot formula about a kinder and gentler world. A woman will be there someday, and if she is not the right one, it won't make any difference.

Except as receivers of massive government aid, farms based solely on the male principles of machine use and specialization don't work very well. And like Hasselstrom's bulls, when things get tense, we tend to push and bang our heads into some immovable object. We badly need the female ability to see "widely."

The reason is that nature and biology are not machines; they are not even very machine-like. Farms won't work like the rest of industry. The idea is impossible. We can't fix what is wrong with the whole by fiddling with the parts, because the parts keep changing right there in our hands! Women who have not completely sold themselves to the dominant structure understand this instinctively. They seem to know that success is made of much more than money, that things are more apt to come out right if we kind of gently herd those things along in the right general direction instead of trying to nail them down tight. That "winning" and "losing" are temporary human constructs that don't have a lot to do with the reality of life on earth.

Farms are basic to civilization and important for, among other things, providing a society with its connection to the nature in which it exists. Wouldn't it be a kick if we learned how to operate them in a way that could serve as a guide to other aspects of our lives together?

Farms need both the feminine and the masculine principles encapsulated in an extended family of several generations. After all, nature and biology run that way. Can we realistically expect to manipulate them (operate a farm) with only the male half of our human capacity?

CONVERSATIONS WITH THE LAND

HER LAND

In H. Elaine Lindgren's 1996 book, *Land In Her Own Name*, there is a picture of the two-story North Dakota house built in 1903 by Mary Belle Hanson and her new husband, Olof Pierson. The house is how they proved up their claims on a quarter-section of land apiece. The caption says it was built straddling the line between the two claims, with the kitchen on Mary Belle's acreage, and the remainder on Olof's. This probably says something about how much they trusted the government not to tinker with the residency requirements for homesteading, and it certainly says something about the division of labor in the new household.

Later in the text, Mary Belle's responsibilities are described. She provided food, housing, and clothing for five to eight men in the winter and for fifteen to forty-five men in the spring and fall. This was in addition to her seven children and husband. The book says she sewed, baked, cooked, churned butter, raised poultry, gardened and cleaned. She and her husband accumulated substantial landholdings.

It was my great good fortune to grow up some forty years ago surrounded by women who knew how to cook. From my mother to my aunts to many of my neighbors and mothers of my friends, they knew the uses of a kitchen and stove. And though they may have tired of the chore (I hardly see how they could not have), I doubt they ever questioned its value. People needed to be fed, and it was their job to do it. And so they did. The food they created was simple, tasty, hot and plentiful. And there was generally not any lack of appetite around those tables.

That time of my mother was about the end of that practice and the start of something new. The something new was an ongoing effort by companies who saw large, money-making opportunities in various forms of quick and counterfeit food to convince us that we were too good to do something as old-fashioned as cooking. Far better, we were told, to leave the cooking to the food industry, and use our time to do something more valuable, such as operating the computers that fill our mailboxes with junk on a daily basis.

It has to be true that the women who encouraged their daughters to give up cooking were never properly appreciated for their cooking by the men in their lives. More is the pity. And it is sure to be true that there have always been men who were competent cooks. Certainly there is no good reason for male incompetence in the kitchen today. The difficulty is that we have all lost competence. Not so much for special meals, because many know how to do that and are sometimes well paid for the ability, but for the daily routine of preparing good and simple food.

Perhaps what we need is an evening class taught by older women. It could be titled "How To Do It Without Cream Of Mushroom Soup." The first session could be called "Stay Home In The Evening Once In Awhile."

To anyone who questions the habit of automatically blaming the government for life's troubles, it is apparent that many Americans' lack of willingness to do for themselves has much to do with why things are as they are. For example, our increasing unwillingness to provide ourselves with any of our own food from garden, farm and kitchen has handed the multinational grain companies — which also own most of the meatpacking industry — the incredible power they possess over our government and the world as a whole. It is possible to draw a line directly from the way we live, to the companies that own the seed, feed and livestock that eat that feed, to the processing industries and the retail stores and eateries that sell it all as a package. This is a disturbing thought, hardly one that would have been envisioned by those ancestors who settled and organized our Land of the Free and Home of the Brave.

Fortunately, we have examples to follow and a new structure of farming and eating just waiting to be built. Many of the examples can be found in books like *Land In Her Own Name*, as well as among the older folks — women mostly — of our own families. There are farms that are ready and able to supply food to people who care about food enough to be choosy about what they put on their own table. The drive should come from that hollow feeling many of us have about the way we are living. When we change ourselves, we change the world.

CONVERSATIONS WITH THE LAND

HUSBANDRY

Animal husbandry is one of the pursuits that have been disallowed for modern humans by the technological powers that be. "We'll git the Mexicans to do that," it is said. Modern farmworkers are as apt to be brown-skinned as black, as can be plainly seen on any of our huge "farms." There may be some history of illegal entry that the empire-building "farmer" may use to enforce compliant behavior. Al Gore, who ran for the presidency in 2000, thought agriculture itself should be outsourced. Thus the livestock man joins the craftsman, the poet, the eccentric and the matriarch in the pile labeled "no longer needed."

Going vegan is the current one-hundred percent solution. According to this view, livestock breeding and production ought to be phased out. Few people who hold this belief have bothered to think seriously about the implications and necessities involved with doing agriculture in a sustained way without livestock — or about any other agricultural question, for that matter. And until I see an honest comparison, for instance, between the soil erosion rates in our pastoral livestock production system here in Minnesota and those of a spinach ponderosa in California, I do not intend to take any of these intellectual games seriously. What I do take to heart is the fact that the views of working people never get considered. We never get asked.

We are not cowboys. Most of us who are any good at all at what we do admit that we know a half-dozen women who would be better at our occupation if they saw a reason to apply themselves to it. But women have been tractored out of agriculture. When the technology came in, they mostly found better things to do — at least those who did not find themselves stuck with supporting a hopeless farm by working a job in town. The smartest of us old fools still farming love the fact that the new truck gardening/community-supported agriculture that is growing so fast includes a goodly portion of females, who are better at marketing and do not have to battle testosterone to keep themselves focused upon the necessary job of production. And a woman, when she goes to care for livestock, starts with a better understanding of a farrowing sow or a cow heavy with calf.

Oh, they have tried to surround the livestock with technology to take the personality out of production and the husbandry out of the farmer. And it works in its way. But it is a production system no one wants to think about, and it produces meat and milk that many people do not want.

Some of us have gone back to practices that others of us never left. These farm practices work by means of a connection between the animal mind and the human, by a kind of empathy or sympathy and a mutual respect. The essence of it is conveyed in the old stockman's saying "slower is faster," which we mutter to ourselves as we go to wean and vaccinate pigs or sort cattle for breeding or to handle bulls.

Farming this way is to deal with animal behavior and biology, along with the entire complexity of nature. Things change right in your hands as you are trying to put them together. As you age in the work, you begin to wonder if that change, which sometimes so balks and frustrates, is being created by your own presence in the system. It gets more difficult to see — much less honor — the separation between the biological, natural world and the human.

Animal husbandry may be the human activity farthest away from the "parks and recreation" approach to nature common among most Americans who live completely within the confines of a man-made environment. It recognizes that struggle and compromise are part of any relationship with the natural world — every bit as much as reverence and awe. Patience is the only possible human approach to dealing with something as intractable as a sow with piglets or a bull too long separate from the cowherd. But rather than honor this slow and difficult development of patience (formerly thought to be a virtue) in a certain number of us, and ensuring that it continues down the generations by making sure that it is financially rewarded, we have as a people opted for the technological solution.

While many can swallow the development of biotech corn and soybean crops and their adaptation to huge systems of mechanical production, they draw the line at a similar approach to animal production. Blasting the DNA of a fungus into a corn seed (which is, after all, only going to be grown for livestock feed) seems not so bad, but injecting cows with an artificial growth hormone and then milking them in continuous confinement while their skeletons outgrow the capacity of their bodies to furnish support is another matter. Now we want it done differently.

But it is all of a piece. We will honor farming, or we will not. If we will not, we will not have it. We will have instead what we have been getting.

There remain a certain number of Old McDonalds in the rural areas who would like to produce animals through careful, lifelong attention to genetics of the herd instead of by following the latest fad. Who believe in husbandry and excellent animal care and management. Who have depended upon the "stockman's eye" for their livings. I am one of those.

However, McDonald wants to be paid. He (or she) is getting kind of tired of sucking hind teat. So please choose your food carefully, and then be willing to pay for it.

WIVES, HUSBANDS, CHILDREN

BOYS

We are not doing so well with our boys. I know this because I used to be one. Statistics say that boys are twice as likely as girls to suffer and die from physical abuse. They are four times as likely as girls to commit suicide. Learning-disabled boys outnumber girls two-to-one.

Simple observation tells us that most boys reach manhood able to express one emotion only, that being anger. Half of all marriages fail, and in far too many of those failed marriages the man walks away from the children. Our incarceration rates are approaching seven per thousand of population, up from a mere one per thousand just thirty years ago. The large majority of prisoners are male. Prison building is our other growth industry to go with the construction of suburban McMansions. We have a big problem.

Michael Gurian says in his book *The Wonder of Boys* that boys are tribal. They need to work together in groups. They need to have a leader and be given the opportunity to lead, and they need to sacrifice for the good of the group. This seems right to me.

He goes on to say that they need three families, loosely defined as the nuclear (parents, primary caregivers, brothers and sisters) extended (aunts and uncles, grandparents, trusted teachers and friends) and the community (neighborhood, culture, church groups, government). It is obvious to any careful observer that our modern life, with its extreme emphasis on economic "success" and money, has pretty much destroyed the first, scattered the second, and twisted many elements of the third group into something that cannot be much good to anyone.

Gurian thinks we can glue these three "families" back together with the broken pieces lying around. To his credit, he knows this repair job cannot take place unless we back away from our obsession with financial success. But he goes no further in exploring what that changed economy might look like, and what effects it would have on our lives together. He assumes that someone else, experts in the financial field I suppose, will see to that.

Well, if we look around, we will see there are no available experts. Only us. We shall have to take up the effort ourselves and see what we can make of it.

CONVERSATIONS WITH THE LAND

For me, it can only start with my own growing up, a subject about which I have thought long and hard. When I grew up on it in the '50s and '60s this farm was in the midst of transitioning from a traditional farm to a modern one. But the transition was early enough in the process that I still got to learn to work in the company of older men. I suffered all that raucous teasing, joke playing at my expense, abrupt ordering around and expectations that I would grow up enough to hold up my end that go with a boy doing vital work with uncles, grandfathers, fathers, neighbors and hired hands.

Little did I know this life was disappearing. I remember the growing shock as I got away from the farm and began encountering first a few, and then later a flood, of young men who not only didn't know what their fathers did for a living, but didn't much respect them for whatever they did do. Far too many of them were unable to respect themselves.

It is crucial that boys' work with the older generations of males have economic value. My interest here is in how we begin to get to where we need to go from where we are.

Our usual way of dealing with the past is simply to dismiss it as sappy nostalgia. But we are well beyond sappiness in the case where we had something that worked, even for a minority, and we refuse to revisit it or take up the effort of reconstructing it because it is anti-modern or off-trend. We are into the careless disposal of young people simply because we are too mentally and emotionally lazy to care. It ought to be obvious by now that the reduction of human meaning to the economic and the financial — those principles of which the modern society is so proud — is one of the most destructive forces we have ever unleashed upon ourselves.

The key insight is that we best bring about the future we want by living as if we were in that future now. I assume an agriculture that causes its community to shrink — that degrades its environment, dilutes the quality of its food production, destroys the families who farm — is itself a failure, no matter how rich a certain few farm operators get to be.

So a farmer who cares about this tremendous loss, this waste of boys, will need to take steps to see that what he does control moves him and, by extension all of us, in a good direction. Farmers control businesses, and it is a common assumption in business that control of the health of the business comes in the expense column through careful examination and change. Just as the money spent as the result of careful judgment has a tremendous impact on the bottom line, it also affects the mental and emotional health of those concerned with the business, the impact upon the environment and, I would suggest, the future for all people connected with that business, however loosely.

When a farmer looks at the purchase of any new technology, for instance, he would not often think that by doing so he is lining the pockets of the global economy, and that this is money leaving the community. If he finds another way to get some of the work done, such as hiring a custom operator, changing farming practices to make heavy horsepower less necessary, or altering methods of animal production, he may be able to change the farm toward a more people-, community- and — especially — kid-friendly operation.

We have long assumed that the move toward the future is a move toward technology. It is only lately that we have begun to see this also means a move away from on-farm management, and toward centralized control. Now we must see it in a larger context as a move away from people as well, and especially the people that matter most: our neighbors, friends and offspring.

The most important lines in the Schedule F are about custom hire, hired labor and the support of the farm's people. It is in these areas we can most improve our farms' financial health and the health of society at large, affected as they are by everything we do on our farms. We can only save our boys — and girls, too, I suspect — by needing them. It is all about the choices we make.

CONVERSATIONS WITH THE LAND

CHILDREN

Wendell Berry, who is one of the few people who writes with any intelligence about the difficulties of living a human life in the technological world we have created, has this to say about the family:

"I do not believe that employment outside the home is as valuable or important or satisfying as employment at home, for either men or women... children need an ordinary daily association with both parents. They need to see their parents at work; they need at first to play at the work they see their parents doing, and then they need to work with their parents."

That forthright statement of what is and is not important is about as different as can be from the usual gobbledygook we get from our modern sages regarding the problems we seem to constantly have with our "younger generation." "Quality time" and "flex time" and all our other fancy justifications for fitting ourselves and our children around the overheated economic engine that we worship get short shrift from Berry. People need to work at home, he says. Children need to play at work, and then children need to do the work. Simple as that. And nary a mention of extracurricular activities.

When I got to the University of Minnesota four decades ago, one of the things that shocked me as I got to know some of my peers was what little respect they had for their parents. They didn't know their parents, and resented the control they exerted over them and life as a whole. They didn't have a clue as to where or how to break out of it or, for that matter, how a decent life might be lived. They had no guide. It was what came to be known as the Sixties Generation.

I didn't know at the time that the background from which I came, where a kid might disagree with or argue with his parents but at least he had a basic understanding of what made them tick because he had grown up seeing them live, was on its way out. The experience of my mostly suburban peers — that of absent parents, plenty of money and the TV as guardian — would soon also be the norm on the farms.

All of the elements that went into creating the Sixties Generation are still with us. Nothing has been solved. My generation has repeated old mistakes.

It would be good for any young parent, as well as us older types surrounding them, to consider this well, for our situation worsens daily. Teen suicides, chemical abuse and eating disorders are epidemic. Children are abandoned, mistreated and neglected at casino doorways. Marriages don't hold; violence spreads like a cancer. Carelessness rules the day. There is nothing easy here, for taking some of Berry's advice would be to change to entire structure of our economy and culture. Children do not play at work and then work on today's farms, as advances in farming technology have made that far too dangerous. The machinery, chemicals and even just the vastness of modern farms have made it impossible for children to be anything other than a nuisance. They don't play at work anywhere else either, for the small shops and local businesses that made daily association with parents possible are mostly gone, along with the local life they supported. Most of us now work in a global business to get money to buy from a global business. We could as well live in it, too, and many do so through television and the internet.

We must face the difficult fact that we cannot give our children a better upbringing until we stop separating them from the important matters of life. One of those is work — work that is done because there is a need for it, and done by people who at least some of the time enjoy the doing of it. Work viewed as a dirty necessity to get money is destructive. Work done because a human wants and needs to do it is an uplifting and humanizing thing.

We will have to insist that more our work is done for this second reason to get our children connected again with the world in which they live. It also will need to be done in such a way that it increasingly includes the children.

This will be difficult, as it means breaking and remaking most of our rules about work, life and economics. The task is so difficult that it is certain never to be done by any expert at any university or think tank. It can only be done by ordinary people on ordinary farms and in ordinary jobs who have had enough, and are willing to change.

Our kids need to hear: "Play over there, boy, or you're going to get hurt." And then: "Bear up your end, son. Didn't you eat your Wheaties today?" And finally, "You're giving me a run for my money. You're all right."

And then they will be. And so will we.

CONVERSATIONS WITH THE LAND

COLLEGE

We disrespect the kids who choose to stay among us, and this is one of the primary reasons we are in such tough shape economically in rural America. Any kid who chooses to stay in a rural area, as well as any who moves back, is automatically suspected of lack of ambition, if not mental ability.

Paul Gruchow, a writer born in rural Minnesota, tells of a group in his town that once came asking for his wife, an attorney, because they wanted a boy they thought wrongfully accused to be well defended in court. It soon became apparent to Paul that they wanted his wife not to represent the boy, but to recommend a real lawyer, one from the Twin Cities. The situation with Gruchow's wife was that even though she was a trained lawyer, she lived and worked in the country, so she couldn't possibly measure up.

Part of it is that we hold a college degree in high esteem. Ask a high school guidance counselor how many seniors are not going to college; they will number in the single digits. Why should this be? Many of the jobs that pay the best do not require a college degree.

And we have told our children for at least three or four generations now that they needed to go away, graduate from college and succeed in life (make lots of money) without ever coming back.

There may have been a time when a college education at least meant that the graduate could read something that would pass for literature, could write a meaningful sentence, had learned something about the chemistry and biology of the world he/she lived in, understood something about a few other places in the world including another language, and knew something of the history of at least his/her own people.

Today? Maybe, but the degrees are in things like computer design and electronic sales and public relations. A college education can still be had, of course. However, if you are interested you are going to have to sort through a lot of fluff to find it.

We disrespect ourselves. We have done so for generations, and it is getting worse. It is easy to understand how that would happen, as the world leads us into it. Every time we go into a grocery store and see that every single thing we produce on our farms is cheaper than it was thirty years ago, our own lives are devalued. Of course a few items may cost more dollars today, but today's dollar is a lot cheaper than the 1980 version.

But this is really no excuse. The food is cheap because we have all suffered for it in rural America and because what profit there is in it any more is being taken by chemical and seed and banking companies. What we need to do is to begin the long, painful process of learning to value what we do again. To do that, we need to put value on food — our product — that is not limited to its dollar value. If we can be clear in our understanding of what went into the food we grow — how much of our own work and risk and sense of values — we will find it easier to begin paying for food what it is worth.

Or at least whenever we can be sure that one of our own benefits, as I certainly am not interested in lining the pockets of the grocery industry. But it is vital that we think clearly. I read recently that re-heatable bacon at twenty-one dollars a pound in the grocery store ought to be priced that high because some food company took a risk on it. If this is so, it ought to go without saying that the farmer who raised the pig and brought all his understanding and experience and work to bear on that task is entitled to such a profit!

A sound economy for our rural areas comes from having a certain number of people who have decided to make things better right where they are, rather than pursuing some kind of golden future off in the distance. And a good part of that "making things better" will come from making our own products better or higher in price because they have had value added to them. Value added, that is, by us!

If we can believe in some of our kids, maybe they can help us figure out how to get the hands of the thief out of our pockets, and start getting for the products of our work and risk what they are worth. This requires something of an alternate American dream, one that has to do with people living a settled life in a place that feels like home, and expecting to be paid for the good work they do. If we can learn to value ourselves and our children, we may value our lives, homes and products. It seems like it may all come as a package.

CONVERSATIONS WITH THE LAND

ADULT CHILDREN

For the last ten years we have attempted to manage our farm and business by the principles of Holistic Management. The focus on goals has been a new direction for us. In addition to profitable production they include quality of life, which implies family and community as well as care for the environment. Particularly exciting is the insight that when goals are too financial and too short term, the land suffers in a way that will, among other impacts, destroy profit and prosperity in the long term. Good farmers have always suspected this, though society does not honor that knowledge, and no other system of management I know of teaches it.

I have not wanted to say it, but the principles of Holistic Management are incomplete. The system deals with time in an indirect way, as it teaches enterprise choice and manipulation of land and animals to achieve goals, most directly when discussing grazing and how grasses and grazing animals interact. But as a tool, Holistic Management does not put time at the level of capital except in the sense that use of labor is implied. We have thought that our farm needs to construct a time budget each year just as we do a capital budget, and give it the same power to discipline our management. Like all farmers, we have inherited certain baggage from the past, part of which is the idea that we have all the time in the world, and can always "make" more.

Beyond the idea of time as an expendable resource for use in our businesses, we find as we age a mounting need of time for ourselves; time to do, if you will, "nothing." Do we live to farm, or farm to live? Can we age into seeing the future in the lives of our children and grandchildren? If we can, we need to teach what we have learned in our lives to our children. We can tell them about the hugeness of the universe and the insignificance of us. We can then tell them what they most want to hear, which is that all we have in this huge, impersonal world is each other, and that not forever, but that what we do have can be enough.

Yet there is a spiritual hole in we who are farming now. We can't tell our children these things until we deal with our own unwillingness to admit we have limited time in this life, and that time is hugely more important than money. To fix what is wrong with us, we need our children.

WIVES, HUSBANDS, CHILDREN

The young are better at living in time. Sure, they also live far too much in the "machine," depending on technology for kicks and life support. Where should we, who never saw a tractor or a pickup we didn't like, suppose they learned these things? And too many of them have no real idea about work or the discipline to learn. But they appear to assume, as we do not, that life is to be enjoyed. Many of them live a day in the way a hungry man eats an orange, making sure that every bit of it makes it into the stomach. We, in contrast, tend to get through each day by yearning for the next, or for the end of the year, or for when the fence is finally finished.

This idea took root in my thinking as I considered my life surrounded by grandchildren while also helping mother get along in a severely constricted life not of her own choosing. It was then that I really saw for the first time that business is never really finished, and that business goals are always illusory. It began to be apparent that every day I could move freely about, doing work I often enjoyed and choosing the food I would eat, was a victory of major proportions.

Some of the best thinkers among us have talked about how we need to romance the children into wanting to farm with us. There is a certain truth to that, but I think that we will not be successful in that until we get better at accepting the humanity of our own children. They are not something to be formulated for success in the economy, or tricked into staying or coming home. There are things they know, somehow, that we do not. When we display the humility to ask them about those things, and to listen to what they say, we can begin to assume the status of people who have lived for a while, and may have seen certain problems and opportunities once or twice before. We can begin to be their senior partners, in addition to being their parents.

CONVERSATIONS WITH THE LAND

CRASH

Representative Marcy Kaptur of Ohio took a thirty-something business reporter gently to task a few weeks ago. Speaking from the floor of the New York Stock Exchange, with a broker at his side, the reporter demanded to know what Congress meant by turning down Treasury Secretary Henry Paulson's demand for a $700 billion bailout of corrupt investment bankers. No doubt seeing the balance in his 401k melting away before his mind's eye, the reporter wanted to impress upon the congresswoman what a high urgency this was.

Kaptur's memory goes back to the days before 401k plans. Her birth at the beginning of the Baby Boom virtually certifies she was raised by people who had direct knowledge of the bad behavior of the rich in the '20s and '30s, and how that behavior cost nearly everyone else dearly. She cautioned the reporter to quiet down and get the panic out of his voice, saying that tone was important now, that we didn't need to add to the panic, and that capable people were working on it.

I hope she was right. I do have to say that the initial congressional refusal of Paulson's stickup looks better than its eventual capitulation in view of the fact that the markets didn't react optimistically in the weeks following the handing over of the cash.

I have often pondered in these weeks about the difference between today's working class and those people of eighty years ago, our grandparents and great grandparents. It is hard to escape the feeling that much of the capability for coping with tough times is gone. This is important to think and talk about, for everything — the suffering, the getting through, the getting by, the choice of radical hate-filled politics or not, the uses of and need for religious grounding — is going to fall on us. That means all of us who have been doing the work of the world all along.

Many of the same facts that have made it so easy for the wealthy classes to snooker the rest of us are going to be a problem in surviving these times. We appear to believe, for instance, that physical work is something to avoid. I don't want to milk those cows, we say. "Let the Mexicans do it," we whisper to each other. And the "masters of the universe" types — those who have just come to our representatives hat in hand and begging for money — hear this. They know they can safely sell our good jobs overseas and shut down our production of useful goods so long as they can figure a con game (credit card debt and sub-prime mortgages) to make us believe we have money to spend.

Family is another thing. Many of the immigrant folks have very tight family structures — a "one-for-all" sort of approach — and a belief system in which the parents understand that when marriage happens and children are created, they have made a commitment to cease living just for themselves, and to take up living in good measure for the children and for the future.

So did we all, at one time. But excessive materialism is an easy master in some ways and we, the somewhat longer-term populations of our country, have let individualism tear our family structures apart in too many cases.

How surprising is it that family doesn't work for us? A look at the television sitcoms and the movies shows a real surplus of hopeless forty-somethings who have not yet figured out how to live in the same universe with their parents. People not surrounded by people they can depend upon tend to buy more of what they have little use for, a fact much celebrated in the corporate board rooms and responsible for the entire industry of advertising.

If family has no value, if it is not there to anchor and support us, we become sitting ducks for a corporate structure that wants to use our work as a "fungible" commodity. Something like corn, moving us around where we can do them the most good, and finally treating us as disposable goods. Family businesses are not started by families who see no value in their members. It must be a given that a family who can start and maintain a business through generations must have strong roots and a sense of place. When family-held businesses are not a large part of any community, community suffers and withers. Families — their businesses, their strong roots in place, and the communities they support — these are the only answer I can see to what is failing on Wall Street.

These are pretty terrifying times, especially for those like me who have no crystal ball. I have no truck for credit default swaps or derivatives, or whatever other names they give their various swindles. To me, it makes no difference if Wall Street ever recovers, though I am pretty sure that with all their resources, including bunches of our money, they will figure out a way back. We can be sure that however much they suffer for what has happened, it will be worse for us in the working class.

But the real economy — the one by which we feed our children and provide our families shelter and build a decent community around us — that one I care about. We are going to have a long, hard road putting back together the precious things I have named. I think it is critical that we formulate wisdom out of this, and that we figure out a way to pass such wisdom on in our children and grandchildren, just as surely as we do our genetics. The wisdom gained from the Great Depression lasted through about two or three generations, during which the working class did pretty well. Since about 1980 it has lost its ability to exert power politically, and lost its self-respect in far too many ways.

Whatever else comes of this situation, it should provide an opportunity for anyone who can do a practical and useful piece of work to be valued again. That would be welcome, indeed.

CONVERSATIONS WITH THE LAND

COMMUNITIES

The law locks up both man and woman
Who steals a goose from off the common.
But lets the greater felon loose
Who steals the common from the goose.

CONVERSATIONS WITH THE LAND

HAY

Hay cutting is thinking time. This dates back to my boyhood on the farm, when we used tractors that were considerably more obnoxious than today's models, and hay mowing was about the only time the tractor was quiet and cool enough to allow thought. Most of them heated and vaporized the gasoline right back at the driver whenever they had to pull anything of a load.

But hay mowing was different. It still is. Maybe I have a strong connection in my mind with some good things that came with the haying, such as working together with neighbors. Whatever the case, there is something about the circling on a relatively quiet tractor, watching the birds and butterflies start up from the hay falling over the cutter bar, and eventually the smell of curing hay, that puts me in a contemplative mood.

As I cut, I thought about what became of the other two fox cubs on the hillside. There had been six at grain-seeding time. Now there were only four. I checked the bone pile around the burrow and saw some lamb legs and a couple of pig bones. I hoped the animals had been dead when the fox found them. I don't trust it.

Up on the other side, the bar kicked up a beer can some one hundred feet from the road, giving rise to a quick flash of anger. Why throw it out? How dare he? I don't understand.

Yesterday, a fellow who stopped up to buy meat gave us a card on which was written this poem, identified only as a medieval quatrain:

"The law locks up both man and woman
who steals the goose from off the common.
But lets the greater felon loose
who steals the common from the goose."

Common or commons, of course, means property in common or commonly held wealth. The middle ages would have understood it to include all public property, all assets that are available to all, and everything that is not properly thought of as personal property, or ordinarily thought of as belonging to God.

COMMUNITIES

The middle ages did this much better than we do. Our great perversion is that we try to make everything into private property, even including life itself; witness the seed patenting efforts of "bioscience" companies. As for our care of our common property in general, ask any janitor at a courthouse or public schoolhouse about that.

The fellow who pitched the beer can was showing a disrespect for my property. He knew if he thought about it (assuming he does think) that the can would probably end up in my field. Even more discouraging is that he shows no respect for the roadway, which is property in common or held by me and him and everyone else in the township, county and state. If he holds no regard for what I own, eventually he will cross an active and aware landowner who will set him straight, perhaps bending his nose in the process. But if he holds so little respect for our property in common, he has no self respect.

If I discover that fox to be helping herself to my lambs and pigs before they have died of some other cause, it may get to the point where I will need to carry the rifle out there and thin the population of foxes to a level I can live with. But if I am fully aware of life in this Creation, mine as well as the fox's, I will do it with a certain caution and regret. There is my need to make a living within the economy we are stuck with, but there is also a need for foxes.

There is no serious philosophy I know of to indicate that I and my farm have any larger ultimate or universal right to be here than does that fox and her cubs. There are only the laws, customs and traditions we have piled up that function pretty much to give us "rights" to do whatever we wish, at whatever the cost.

There is a sense in which that field is a commons for me and my family and livestock and the fox and her cubs, as well as the birds and insects that fly up from the cutter bar and everything else that calls the field home. To think of it as such, whether or not it is or could be, seems to me to be a way of encouraging kindly use of it by all of us. And kindly use is a result good enough that I tend to think that applying the philosophy of the commons is a good idea.

But how in the world do we get any of that idea across to the guy who hasn't yet reached the level where he will dispose of his beer cans in a more responsible manner?

CONVERSATIONS WITH THE LAND

DECLINE

It is possible in some grazing and marketing circles to get a regular lecture full of wisdom from the trendiest corporate gurus — people who write books focusing narrowly upon the golden key or single insight most likely to unplug the most stubborn of business bottlenecks and flood the lucky learner in a veritable sea of cash and success. These keys are generally practices and procedures used successfully by such notables as Donald Trump, Warren Buffet and Bill Gates. They always focus within the context of conventional market economics, and always in the service of the individual's financial well being.

This is common fare throughout our society and economy; certainly it is not limited to grazing or even agriculture. Self-help wealth schemes depend upon the belief that success means money, and only money, and that whatever problems any of us have are always individual problems and amenable to whatever deals the individual can make with the surrounding culture that is thought to be, essentially, the market economy.

But the neighborhood in which I live and farm has changed completely in the nearly sixty years I have known it. Some of the farmsteads exist only as ghosts in the minds of those of us closing in fast on elderly, but many of them are still there. None, however, are farms in any sense of the word. And while I am still surrounded by people on all those yards, I do not know many of them and they do not know me. When they drive past as I go about my business in the pastures, I often do not look up anymore for, when I do, they generally are not looking at me. I do not know what their lives are, or what they are like.

There are locks on the doors of this house that had none when I lived in it at age ten. If we have a fire, the neighbors won't notice, because they are at work. If the cattle escape the fence, neighbors will not call me; they will call the sheriff. This is a human damage for which Donald Trump has no cure.

When I need to talk to someone about planning for the winter's feed, or about the kind of crop rotation with no large corn or bean component, or work scheduling on a diversified farm, I go to the electronic neighborhood, using web sites and e-mail. I am thankful that these networks are available, but there are days I would give anything to lean on my corner post and talk to my flesh and blood neighbor about what he is doing, what I am doing, and what is going on in the community. Shoot, I could even figure out how to talk to a crop-farming neighbor, if I had one. I might find out who died in time for the funeral, something that has been a problem since I decided to do without the local daily paper.

This year, the school my grandchildren attend is back down to seven hundred students, which was the size of the largest of the three schools that went into making it when it was put together by the school board on which I served in the 1980s. The school district is asking for a referendum to keep the doors open, and the board is opening talks with neighboring (also consolidated) districts. I traveled two and then eight miles to school. My children went eight miles, my grandchildren go twelve or fifteen and, if there is another generation on this farm and if there is a school, it looks like my great-grandchildren will travel twenty-five and more. This is more damage in the name of economic progress.

I am not sure if the University of Minnesota is graduating any large-animal veterinarians who plan an independent practice. If it is, there can't be many. Our vet is part of a clinic that was two large animal vets when it started up twenty-five years ago, both of whom did a little pet work in the late afternoons. That practice is now seven vets, with only one-and-a-half who will even come out on a large-animal call. When these two guys quit, we will be down to going around to the hog companies, hat in hand, begging for a little vet service.

My neighborhood has progressed through several stages on its way to extinction. It began as a kind of a frontier, mutual-help affair that produced ad hoc solutions to problems. It passed through a time when folks understood themselves as being sorted out on the basis of success or failure in modernizing their farms, then to a gaggle of "entrepreneurs" competing for a limited land base and limited profits, to where it is today.

Today we are former farmers who drive fuel trucks and stand on assembly lines, leaving behind a few bewildered fellows on farms wondering where everyone went and how the world managed to change so fast. Not only do I not have any neighbors in a position to notice my barn burning at mid-day, I don't have any to notice that we need another barn to employ the two families that work here, and to help us build it. I have never in my life helped a neighbor build a barn, either, which demonstrates a kind of regrettable symmetry in our rural lives, and a marker for how far we have come in the wrong direction.

This is why our goal-setting must always include more than profitability. Not one of these problems can be solved by money, even if the floodgates can be opened as the gurus promise. And not one problem can be solved by any individual. As individuals, we have cut the last deals we can; now we must think a new thought. Our rural communities are necessary to our farms, and they require people. We must learn to think along the lines of the late Minnesota Senator Paul Wellstone, who said, "We do well when everyone does well." The trouble is that it is getting hard to believe, as Wellstone did, in the efficacy of the government in making this happen, for government is more and more in the pockets of the corporations, and under the control of its own CIA spooks and military.

It is going to be up to us. It will take a generation and more. We can begin with admitting that money is not enough by itself. Leading with that thought would be a change.

COMMUNITIES
ALICE

It is difficult to make the argument against irradiation of meat. Anyone who does so is arguing in favor of responsibility on the part of everyone, from the farmer through all the channels of processing and retail to, and including, the end user. Arguments in favor of responsibility don't go well today. We aren't heavy on responsibility. We are heavy on "fun," or maybe "wealth."

And anyone who makes the argument is apt to be accused of thinking the earth is flat. Some state functionary came up with that tired old jibe the other day when a reporter shoved a mike under his nose. Someone really should promote a flat earth society, since I so often find myself labeled as a believer. It could be a poet's society of some sort. We could sit around drinking clean water and watching the horizon.

Suffice it to say that irradiation makes sterile whatever it blasts. What this means is that if the piece of meat was nuked because it was suspected that oh, I don't know, sloppy slaughterhouse practices might have gotten it covered in manure during processing, the nuking will remove the infectious agent, but not the manure that carried it. Dinner, anyone?

Supporters of the concept have had the last word. A reporter showed up at a picnic recently, asked a picnicker what she thought of the irradiated burger she was eating, and was told between mouthfuls, "I dunno. Tastes like a burger." 'Nuf said.

Storytelling is the last refuge of the loser. So you are going to get one. When we moved to St. Paul forty years ago in the early years of our marriage, it was our great good fortune to land in the middle of a closely knit community in the western part of the city. Alice Fleming, the widow who lived upstairs, befriended us and over the years told us about the neighborhood. She was like most of the rest of the older folks around: Irish, Catholic, and railroad connected. She walked with us down the street, pointed to one after another of those wonderful old houses, and told their stories.

They were about the daughter eloping out of this one, and how the lady on the corner lost her husband, a section chief, under the wheels thirty years before. Failed and successful businesses showed up in her talk, good and bad families, the son who was a lawyer and the husband who drank. And always that community was there. It was unthinkable to her that she should be anywhere else.

CONVERSATIONS WITH THE LAND

I remember Alice in the Don's Country Boy grocery store on Fairview and Selby as clearly as if it were yesterday. She bought meat, and this is how it went.

Alice went back to the butcher's counter and, pointing to a round steak or a nice roast, told the man she wanted a half-pound of beef ground from "that." She drew herself up to her full five-feet two-inches, and said, "Grind it even and no gristle! I'm watching you!"

Alice knew that butcher. In fact, I am pretty sure she knew the butcher's father. Certainly she and that same butcher showed up every Sunday morning or Saturday night at St. Mark's church just to the west. She may have shared a pew with him more than once.

Alice expected to instruct the butcher because she grew up with the sure knowledge that she would have to see to herself in the world. And she expected the butcher to expect that kind of instruction. If he was smart he knew that if Alice got a bellyache after eating something he sold her, she was going to come looking for him.

If he had any intelligence at all, when ordering his meat the butcher must have realized he was taking a considerable risk if he did not do his best to ensure he was getting exactly what he wanted, and that it was clean and slick as a whistle. The butcher had, after all, been selling on his personal responsibility to a series of Alice Flemings who came into his store. They knew where he lived. It is not beyond belief that he made it his business to know where his distributor lived, and with whom the distributor dealt.

Nowadays the Country Boy shop is an antique store, and meat is sold at the big box under plastic, which I suppose is reassuring because this seals the outside contamination out, and the inside contamination in. I often wonder what Alice Fleming would have made of meat under plastic. I never saw her buy any. I think she may have insisted on seeing the plastic being put on, and having a conversation with the person doing the work.

A flat earther might wonder what was wrong with that attitude.

COMMUNITIES
PAO YANG

Pao Yang must have been in his forties when I met him in 1998. Compactly and powerfully built, with a ready smile, at five feet and two inches he was a full head shorter than I.

He liked to talk, even though English was his second language. To look me in the eye he would try to maneuver me around in the break room of his old slaughterhouse so that he could stand on the raised portion containing the kitchenette, while I stood a step down. On our first meeting he told me in his rapid-fire, broken English about his boyhood, about his people carrying rifles against the Viet Cong in his homeland of Laos. About spending too many years in a refugee camp before finally being allowed into the U.S. About working the night shift at a downtown Minneapolis laundry to accumulate enough money to marry. And then how the chicken bones said that he had chosen the wrong woman, and how he had to look some more to find the right one, and then married her. How they both worked in the laundry until they had gathered enough money so that he and the rest of his tribe could buy this little, broken-down old slaughterhouse next to the soon-to-close St. Paul Union Stockyards, and begin to supply the kind of meats to which his people were accustomed.

Pao Yang was Hmong. The slumlords in the city of St. Paul were not terribly fond of the Hmong because they tended to haul soil up into their apartments so they could grow food. I only spent a few hours with Pao Yang in his slaughterhouse, but I saw a lot there. They de-bristled the hogs instead of skinning. People came to the plant and chose the live pig they wanted out of the pens in the back. The pigs were always at least somewhat undersized. After picking their pig, they helped with the slaughter, cleaned up intestines, cut up stomachs, made sausage.

The place would be crowded with people and noise, kids running and knives flying everywhere. Families came and worked together. The establishment was owned by seven or eight brothers, as nearly as I could tell. Those times when I heard the stories from Pao Yang were times when we were waiting for his cousin's son or his nephew's girl or whoever understood writing well enough to write my check. Hmong is spoken, not written, so those young people had already come a long way. The checks never bounced.

These were times when I was regularly feeling pretty sorry for myself. The markets had driven the hogs down to about eight cents per pound that fall, and we had many pigs on feed. I had found Pao Yang and his need for undersized pigs in order to try to get something for the hogs we had on hand instead of just shooting them to stop the feed bill. I made several trips that fall, the last few in the company of two other farmers hauling our pigs. The price paid for the pigs just covered the use of their trailers. My trips weren't easy in November either: 120 miles one way in an old Power Wagon with an iffy heater and pretty casual brakes.

But then there were these words in St. Paul, from a man in his thirties. "I had two beautiful water buffalo in Laos," he told me with a faraway look in his eye. "In the summers, they would graze in the high country and then I would go get them when there was work to do on the farm. There I always had the food I needed. If there was to be a celebration, I would just slaughter something I already had. But here, when my son becomes a man, I must come here and buy a pig to celebrate with my family and friends."

Tight corners create gumption. The older I get, the more convinced I am of this. Farming had ground me down by November of '98. It was, I thought, about as bad as it could get. And then, as if by accident, I spent a few hours surrounded by people who came out of tough circumstances I could barely imagine. It was this experience as much as any careful thought or financial analysis that started the process of turning this farm around. The thought of being totally uprooted in young adulthood and having to start over in another world trying to take care of family brought me, in late middle age, to being able to consider a different form of production and marketing that eventually brought the farm to where it could possibly become secure in the hog business.

These thoughts come to the surface again all these years later as I am trying to make sense of all the anger and violent intent I hear around our country as tough economic times grip us ever tighter. White American males seem particularly prone to it. It is white males who carry guns to health care hearings and casually threaten others. It is whites who want to "take the country back" — from whom they never quite say.

What I had to face in 1998 was that what I felt had been taken away from me by American agriculture was something I was probably not entitled to expect in the first place. The world is not made that way. Plenty of people have worked harder, risked more, and lost out in more important ways than had I.

COMMUNITIES

Some of them were short, brown people who had simply chosen the wrong side in a war.

Now, over a decade later, I have no answers to the problem of who wins and who suffers. But I do think that the proper response to the constantly increasing pressure the global financial elites, with their economic and military might, put on ordinary people and their communities is for all of us to learn about making common cause with others who may look a little different.

Dispossession affects all of us. Just as it is wrong that farmers are driven from their land in the United States, it is right that people like the Hmong have their chance to farm. It is all one big problem — one for which a solution cannot be bought, but must be made.

CONVERSATIONS WITH THE LAND

LOCAL

There are some fringy nutty types running around who seem to think they know a thing or two about what would fix the schools. Here's a favorite of mine, told to me by a teacher I admire, about how to catch our children up with the Japanese.

"First," he said, "cancel all sports. Then disallow all part-time jobs. Increase homework. Kids should be expected to go home and study and do family chores 'til supper, then study some more and get to bed at a reasonable time so that they come to school the next day rested and ready to learn."

Bear in mind that the news this fall is that the experts want schools to start a little later so that the teenagers who worked until one o'clock the night before don't have such a hard time staying awake. Also in the news is the fact that many schools are having trouble finding enough bus drivers, possibly partly because many people would rather not have to cope with the kids.

In the event that the schools ever get themselves delivered of the necessity of being babysitters and/or semi-pro sports franchises, I have a few suggestions that I think might improve education. They center around the importance of the family and the place. I want to say right up front that this is not a recent failing of the schools, though I don't believe the situation has gotten any better while I have been aging these past four or five decades.

I had to graduate from my high school, then from four years of Minnesota higher education, and then put in ten or more years within the workaday world before I began to be aware that there were capable writers and storytellers in the very area I was born. While it was with good reason that I studied as part of my English education people such as Shakespeare, Chaucer, Milton, Hawthorne and Twain, I had to discover for myself people such as Robert and Carol Bly, Frederick Mannfred, Wil Weaver, Linda Hasselstrom, Paul Gruchow and Bill Holm — all rural Minnesota and Dakota writers, and all with plenty to say about who we are and why we are that way.

The situation in literature is an example of the education we received and our kids are still receiving in all areas. When you think about it, it is no wonder they're leaving. They are learning that everything interesting and worthwhile is somewhere else.

COMMUNITIES

Do our high school graduates leave knowing anything about the Sioux uprising of 1862? Do they know as much about it as they do of the great Chicago fire? Do they know why it happened, and some of the results?

Can any of our children name our local soils? Do they know the difference between our prairie mollisols and, for instance, the karst soils of southeastern Minnesota, so that they might have an idea of what each allows and forbids in agriculture? Do they know where the farm drainage goes for the local area? Do they have a clue as to our agricultural economics? Do they know anything at all about how agriculture has changed in the past five decades, and what kind of impact those changes have had on their lives, including their school?

Have our kids any idea of their own family and ethnic history, such as why great-grandparents moved here and what they found? Have they heard or read some of the firsthand accounts of the first white people to see this place? Can they say why their town is different in some ways? Do they know what the main business of the community is, and if there has been any change in that over the last century?

How many local species of trees can our students name? Do they know what is native here, and what is a carefully cultivated transplant? Do they know the difference between a warm season and cool season grass? Can they point out and name four different birds?

The argument will be made that students are not taught these things because they must be prepared to live and work in a global economy and culture. It will be said that the Minnesota prairie is unimportant and of no interest to anyone significant, that they better prepare themselves to leave, and the sooner and the farther, the better.

But it is not possible to live well in a place while considering it to be unimportant. To try to do so is to live by gazing into the TV set. And it is not possible to live well and work in a global culture. Living well and working well can only be done locally for good and personal reasons. If we take care of our local place, the globe will be well taken care of. The first step is learning how to take care. And that starts with learning about our place.

CONVERSATIONS WITH THE LAND

WRONG STORY

We tell each other the wrong story here in farm country. In fact, we tell each other several wrong stories that, taken together, go far to explain our economic and social predicament. I first became aware of this tendency a few years ago when my family needed to sell land so that ownership for the next generation and the financial health of the farm and its families would be safeguarded.

We had decided in the early 1990s that continuing to operate part of the family holdings five miles distant from the main farm was going to be awkward as we switched operations away from cash cropping and toward land-based livestock. So the land was rented out, and we set about building a farm business that would operate without whatever cropping income would have come from continued operations there. When the family sold that parcel to settle the estate a full ten years later in the early 2000s, the sale caused no pain to the farming operations.

At the time, the land market here was such that there was a gap of perhaps several hundred dollars per acre between what a local operating farmer would or could pay versus the offer of an off-farm city investor who would buy it and then rent it out at top dollar. We chose to sell it to a local farm family, one that had been in business here for several generations.

We were immediately the main topic in the local coffee shop, I guess. I found this out when several weeks later a neighbor whose son was taking over his farm began commiserating with me over my not having waited quite long enough to hear of the great price being paid by the Twin Cities group. I smiled and shrugged.

I regret that, thinking of it now. I should have responded with something angrier, more to the point. Some people need that, especially those who spend part of every morning sitting at the diner in town, making heroes of those with the most money and the greediest grab.

We farmers are not alone in this, as all Americans are part of the craziness. Wall Street died of its own decay. It has no useful purpose that I can see. That our government chooses to keep the corpse moving does not change the fact that Wall Street's major idea — bottom-line thinking — does not work.

And if it doesn't work on their "Street" that runs on so much of our money, it certainly doesn't work on ours. We pretty much cannot think another thought after so many years of being told that markets solve everything, to the point that markets even make our moral choices. But think we must.

We should start with the principle that a sole focus upon money always causes moral decay. If we can internalize that, which is so close to what so many of us say we believe in our religions and churches, then I think we can start to look wider in our efforts to put together a reasoned and reasonable approach to life on earth.

Focusing only upon profit does not keep schools open and available for rural children, including the grandchildren of the neighbor named above. Ordinary human beings on school boards do that, sometimes sitting until two in the morning trying to balance the budgets, cooperate with neighboring schools and cope with the state government.

Profit-driven farms do not worry much about who will farm after them, or how many farms the farming country needs. For instance, the only two possible responses of a profit-only farm to the coming shortage of independent large animal veterinarians are to either get big enough to have one on staff, or quit livestock altogether.

The question of available feed mills gets the same answer. In the profit-only view, farm supplies should always be purchased in quantity and as far away from the farm as is necessary to get the best price. The fact that the local lumberyard closes up and the owner moves his two children out of the school is not the farmer's concern.

This kind of thinking gets the kind of political representation it can buy, of course. The representative of the rural district will think he has served his district well if he can keep huge checks flowing to the six or eight largest and wealthiest crop farms in each county, regardless of the impact upon the community or agriculture as a whole. The state legislatures will run mostly to the benefit of the land developers under the same "throw the money in at the top" thinking. Local governments will bring exploitative, low-pay, short-term business into the community with the promise of tax forgiveness and help with the capital investment, paying no attention to local farms and business startups that lie beyond the reach of these goodwill baskets.

The only cure for our current bad stories is a better one, constantly told and consistently held to. It is something we can do only for ourselves; the TV will not tell it, neither will the internet. It will not come from above us in the governmental and business worlds, and whatever does come from above will not benefit us. No one there thinks there is a problem, or cares.

I had a friend who died suddenly a few years ago in his eighties. In what turned out to be his last year of life, I went over to pick raspberries with him, even though we had a surplus here, because he was grieved with the thought that due to reduced appetite in his own house, the birds would get too many. He tended his gardens and ran his farm that last year of his life. And he planted trees to fill in a gap or two in the field windbreaks.

When the memory of my friend and the current lives of others like him are celebrated in the coffee shops instead of the latest land grab, then we will have achieved something real — something the powers that be don't want us thinking about. And in the process, we will have started to become the kind of community we want to protect and build upon.

COMMUNITIES
GOOSE

"The law locks up both man and woman
Who steals a goose from off the common.
But lets the greater felon loose,
Who steals the common from the goose."

— Medieval quatrain

We have essentially no "common" left in our culture. This is the cause of considerable applause on the political right wing, which believes strongly in the virtue of private property and entrepreneurship.

Witness Minnesota's Tim Pawlenty, who after eight years as a governor dedicated to the slow starvation of both public education and public health — not to mention the transportation system here in my state — has decided to inflict himself on the rest of the country by attempting to move into the White House. Pawlenty says that if a good or service shows up on Google, the government shouldn't be doing it.

There is, of course, much reason to be suspicious of everything the government is doing now, and that suspicion comes from all quarters of the political spectrum. But many of us also believe that government has an important role to play in securing some semblance of the "common" in our lives, such as reasonable security in old age, access to affordable medical care, education for children, fair treatment at work, protection of the weak and so forth. Evidently that belief is no longer shared completely across the political spectrum.

But it is quite possible that someone who disagrees with me in the matter of Pawlenty and the right wing could wholeheartedly agree with my desire for some kind of return to the commons. And that is because the idea of the "common" is at best an uneasy partner with the concept of government involvement in its citizens' lives. It means much more than that.

When Monsanto invents a corn seed that will help it sell its crop chemical, and the crop pollinates promiscuously, contaminating the entire corn seed supply both open and hybrid, it has stolen a piece of the common. That the U.S. Patent Office made the action legal doesn't change the fact that the corporation is enriching itself by interfering with everyone else's access to clean seed from public sources. That Monsanto is planning a Terminator hybrid to produce sterile seeds — and which is just as apt to cross-pollinate with the traditional varieties to produce sterile seeds in all lines — is pretty terrifying to contemplate. It is privatization run amok.

Corporations are being allowed to buy up the Earth's water supplies in the same way they buy the oil reserves. Clean water was generally thought to be a human right, but it is fast becoming a profit opportunity. Why do you think we are being advertised into drinking purchased water out of a plastic bottle? For the same reason we are being coaxed and herded into charter schools: so that we will learn to devalue the common (public) product.

When selling raw milk becomes a criminal act, we must wonder how long it will be before a farmer is jailed for drinking the milk from his own herd. The dairy industry has carefully put together a sophisticated processing, marketing and distribution system that reserves the profit for after-farm business and industry, and it has every interest in making sure every gallon of milk is marketed through its channels. Author/farmer Wendell Berry wondered four decades ago already why it is with us that sanitation must always be expensive.

As you can see, these examples of assaults on the common include instances of government heavy-handedness as well as government failure. That is because the very idea of the common is not about government. "Common," which derives from the common pasturage available to be used by all in medieval communities, is a social custom properly buttressed by a long tradition of peasant preferences for shared work, community life and religion. Much of its foundation undoubtedly came from pagan sources, as at the time most of Europe was not that far removed from paganism. The American natives also lived by holding everything in common, and could not understand land ownership.

The Bible is full of references to this approach to our lives together. Take a look at the Psalms and the Proverbs. Nearly everything Jesus said was closely connected one way or another to the commons idea: "You are your brother's keeper," for instance. The early church ran on a "common" basis.

Our political categories don't satisfy anymore. They don't speak to any close understanding of either human nature or human problems. Every problem I see in the rural areas or with farming or our food economy leads to the same conclusion.

COMMUNITIES

What must change is our behavior — the way we live with each other, the way we do our work, the way we raise our children. New laws or new government programs, well intentioned as some may be, are clumsy tools indeed.

We have a long way to go. Much of what passes for our accepted wisdom is stored in this computer and its word processing program and others like it. Undoubtedly you will not be surprised to learn that the program does not like the word "common" as it is used in the medieval quatrain atop this column. The word is underlined in green, meaning that its use in the quatrain is not ordinary or accepted usage in the sentence containing it. This usage does not please the Silicon Valley hotshot who wrote the program. This person has no basis for understanding the saying, and probably feels superior because he doesn't.

The program doesn't like the second use of the word "goose" either. I suppose this is because it cannot conceive of a world where anything can be stolen from a goose. But I can.

CONVERSATIONS WITH THE LAND

ENABLING THE SCOUNDRELS

ENABLING THE SCOUNDRELS

SLAUGHTERHOUSE

The political air is now full of the need to "get the economy going again" and "jump start Wall Street" and "get our confidence back." And it is easy to understand the need to do something — anything, even if it is wrong — when people are losing their jobs, and with them their ability to care for their children, to plan for illness and retirement, to feel the satisfaction that goes with being at least satisfactorily employed.

But all jobs are not being lost, the stock market has not fallen to zero and, most important, the number of hungry people has not yet increased precipitously. Before it does, we ought to remember that like farms, large things such as economies do not make major changes easily. This is especially so if large government handouts are to be had. It seems likely to me at least that there will be no real change until and unless the people operating the economy, which is all of us of in some sense, are forced into a corner where we must either change or lose it all. I am speaking of course of the real economy, by which things are made and the work of the world gets done, and not that collection of delusions we call Wall Street.

As we talk about change in the economy, we should persist in reminding ourselves that we weren't overly fond of some of what we had when we did have it. In a republic, discontent leads to political action leads to policy change. At least that's how I learned it in high school. Here is a list of what's wrong that comes out of my head, in no particular order.

- The economy does not honor productive work. It rewards the con artist and the ad man over anyone with dirt under his fingernails. Work caring for people, such as nursing and home health care, has been particularly scorned.
- It honors the ability to buy over doing, thinking and creating.
- The economy sells out the working class. It has emptied the countryside, destroyed rural communities and cheapened labor nationwide.
- It has made health care a for-profit enterprise in direct contradiction of the medical profession's Hippocratic oath, and then priced it out of our reach.
- The economy is based upon the destruction and waste of resources, including land. Petroleum is wasted moving California oranges to Florida. Water is wasted growing alfalfa in the desert.

CONVERSATIONS WITH THE LAND

How can we build a new economy, one that better cares for the land and that honors the best possibilities within us? How can we do such a thing intentionally when it seems as if the economy we do have grew in place as naturally as a plant? The best answer to the second question is that it did not grow in place naturally, that indeed we planted, fertilized and cared for it.

I spent the first half of the 1970s working at the veterinary clinic on the campus of the University of Minnesota. Part of my job was to supervise the student help doing the work, which consisted of animal care, barn work, haying in the summer, and so forth. My student crew was entirely veterinary students.

I particularly remember one I will call George, who was in his last year of vet school. As with most of his classmates, George's course through to his license involved seven years of university study. But unlike the rest of the students, George's job under my direction was his very first non-summer job in the nearly seven years of university study. This happened because his father insisted upon it.

George got the government grants that were available to all students at that time, but he had no student loans. His tuition, books and all his living expenses were covered by checks from his father. In return, his father insisted that he treat his university education as a career, and concentrate on it so that he would learn as much as possible.

George admired and respected his father, and spoke of him often. His father was a high school graduate who had never attended a day of college classes, and who supported the family and educated his son with his earnings from working in a slaughterhouse. George said his father often worked on the kill floor because it paid a little more.

It is difficult to imagine a modern slaughterhouse worker being in a position to make such a deal with his son. This is because work in the '70s was union work, and it was possible to decently support a family with the earnings. Today it is not. There is a history to trace here, and it shows clearly that in some very real ways, we chose the economy we live in.

Ronald Reagan came to the White House in 1981, and one of his first acts as President was to fire the air traffic controllers who were on strike. This was applauded by many people — even by many in the working class who talked about the need to bring the controllers down a notch or two. Of course, the lack of real reaction to those firings emboldened Reagan and those he brought in with him to begin a series of changes in the National Labor Relations Board,

ENABLING THE SCOUNDRELS

setting it on a tilt away from labor and toward corporate power. This would continue to grow through Bush One, then Clinton, and on to Bush Two. We will see about Obama. Sticking it to working people is now a bipartisan game.

After the controller firings, slaughterhouses were some of the first businesses ducking out of union contracts to hire replacements at not much more than half the union wage. Working conditions deteriorated for the same reason that wages did: because the powers that be could get away with it. Part of our debate over immigration needs to recognize the fact that the government allowed meat companies to drive wages and working conditions down to the point that no one but an immigrant, and often an illegal one, would consider taking that job. And we kept re-electing those folks.

I also notice that it is difficult to talk in my community about the millions provided to the bank executives leaving their positions, plus the extremely high levels of auto executive compensation, without having someone wax indignant about the janitor at the Ford plant in St. Paul who is paid ... I don't know ... maybe $20 an hour for pushing a broom.

We have done better than we are doing. It would be healthy for us indeed to honor and try to foster the possibility of parents with menial or even nasty jobs being able to sponsor their children at universities. For one thing, this might make universities a little more serious in their roles. But we have traveled far from that possibility today, and sometimes the only way ahead is to start by going back, at least to take a look. We don't have to make it all up from scratch. The example cited comes from our own recent past, and is for that reason not at all unthinkable.

Yet we should not automatically accept an example from the past, even a good one, as a blueprint for the future. For instance, in *The Omnivore's Dilemma*, Michael Pollan presents Virginia farmer Joel Salatin's argument that slaughtering animals is not something anyone should do as a sole occupation; that it needs to be shared among more people to dilute the bad effects it may have on the personality and foster a more realistic view of our food system. This seems right to me, though we are far from being able to think in those terms.

We would first need an economy that could provide for that kind of flexibility and creativity. It could grow out of what many of us are doing with our farms as we take on marketing roles. But whatever the structure and shape of the economy we set about building, it is important that we go forward with the resolve to make it to honor people in their work as they strive and dream. Dreams are the stuff of which true prosperity is made.

CONVERSATIONS WITH THE LAND

PAINTER

In 1970, a union painter at the University of Minnesota painted his truck. Because it had given him nothing but grief in the three years since he bought it new, he painted it lemon yellow. It looked like he had done it with a broom. Then he wrote "this truck is a lemon" along each side of the truck in large black letters. That done, he went to the lumberyard and bought two sheets of plywood and some wooden two-by-twos for stakes. These he also painted lemon yellow before mounting the plywood on each side of the truck's box. Then he took more black paint and carefully listed all of the repairs that had gone into the machine in the three years since it was new. This totally covered both sheets. His plan was to drive the truck to work and back for as long as it ran, and then to give it to the junkyard.

The university was mortified. Perhaps it had some research sponsored by Chrysler, or hoped to. They first talked directly to the painter and, when that failed, talked to his union. The painter told the university that if they planned to pay him from the time he left his front door each morning until he returned home in the evening, he would clean up his truck. Otherwise, the painter said, you can go to hell. The union backed him up. As far as I know, he finished his entire working life as a painter at the University of Minnesota.

I am thinking of this painter again because I have been trying to understand the Tea Party people. The painter was several years older than I and would be retired by now. Would he be in the Tea Party? He couldn't have known that in 1970 when he painted his truck, and throughout 1971 as he wrangled with the university, we were at the very peak of good times for working people in this country, at least the white ones. Things had never before been so good for the working class, and would not be again for at least forty years and maybe forever. What does an entire life of downhill do to a person's frame of mind? What would he think as his kids either failed to find work as good as his, or graduated into the executive and professional classes to enrich themselves by exploiting people such as their father?

It is important, I think, that the system in place for this painter in 1970 gave him a good wage and benefits, plus the opportunity to vent his spleen in a very public way over having been taken in a deal on a vehicle, without putting his family's livelihood on the line. That certainly would not be the case today.

And today, when it seems the best the working people of the country can hope for is some kind of a nitwit in the White House who can "triangulate," it is difficult to focus on the value of the kind of job security that allows for some self expression.

It is not enough that the wages are good, though that would be a step in the right direction, or that the job be relatively secure, which will not happen until we get some control of our corporate elite. There needs to be space for human expression. There needs to be room for a human to live, in the real sense of that word.

The Tea Party is overwhelmingly white, mostly retired, and just a bit wealthier than the average working class stiff. An argument can be made that they are angry that their chance to really live in their work and in their country has been taken away.

For instance, a slaughterhouse worker who may have finished his career in the '90s at a lower wage than the one he started at in the '60s, and who spent the past two or three decades watching Hispanics do the job he did while being paid less than half the salary and no benefits, may well hold up a sign warning the government to keep its hands off "my Medicare." He has lost enough, and he is scared. He knows that his replacements are not really to blame. He knows that his government won't fix immigration because it does not want to.

What he hasn't yet figured out is that it won't do the fixing because the same folks who own the government also own the slaughterhouse — and the Tea Party itself.

Confused as the Tea Party people may be, they have nothing on Obama and the Democrats. These folks just spent two years pretty much doing business as usual, as they seem to believe that the troubles with the economy started in 2008 rather than 1971, and that they actually fixed something by joining the previous Bush administration in handing the Treasury over to Wall Street. The working class still wonders if it would have done the rest of us very much harm to just let Wall Street choke in its own slime. I know I do. I will bet that retired university painter does, too.

CONVERSATIONS WITH THE LAND

SUPERSTITION

A couple of our primary modern superstitions are that communication is always good, and that fast and easy is best. So we get the computer, which, by making possible the e-mail, puts us into the position of cheering for the most vicious and degenerate kinds of human "thoughts" because they can be simply and easily passed to a large number of people.

I heard a radio interview years ago when this computer world was being mid-wifed into existence by the usual gaggle of boosters and glad-handers. The interviewee was holding forth at considerable length about the absolute hopelessness of anyone who would desire the current postal system, which he referred to as snail mail, when all communications could move so much cheaper and faster via the internet. I wondered then if that meant I was going to have to give up my treasured association with my rural mail carrier and neighbor, a man with whom I shared my first real job and some of my college years. I was best man at his marriage ceremony, and I really did not relish the thought of replacing his regular appearance at my box with electronic blips on a screen.

An absolutist attitude is one of the primary characteristics of the true believer. The man on the radio did not believe or say that e-mail would be a good thing, that it would improve our lives, or that it would serve as a good complement to the already-existing service. No, he said that it would completely replace the postal service, that it was better in every way, and that anyone who clung to the postal system would be beside the point and hopelessly out of date.

Our near-religious awe for technologically enhanced communication has to do, I think, with a jumbling of the meanings of such words as information, knowledge and wisdom. We like to pretend that information is knowledge. It is not. Knowledge is, at the least, systematized information — information organized so that it is useful in some way to someone. Knowledge is not wisdom, either, any more than information is knowledge. We would wish it so, because then wisdom would be a kind of quantity thing, something that could be gained by stacking up enough knowledge or systematic information, so that at a certain point the whole pile becomes wisdom and we graduate to a kind of guru status. Wisdom is in reality the engagement of careful and ongoing human thought as well as emotions and memory over time with

ENABLING THE SCOUNDRELS

the hard, rough edges of lived experience. Wisdom is, in part, knowing what to do — and what not to do — with knowledge and information. Wisdom always has a human cost, sometimes a dear one, and can never be bought with a credit card in the computer store.

The kinds of minds that railed against the very idea and existence of the postal system on that radio interview tell us that broadband access will completely cure what ails the rural areas. Now it is impossible to deny that much of what passes for "the economy" these days goes on over the internet. Leaving aside the question of how much of the economy as it is presented to us is real, how much is desirable outside of Wall Street and how much of it is good for us here in rural America, it must still be admitted that slow internet connections in the country do not mesh well with fast ones in the urban areas. It is true also that there are links between the rural and urban economies, thus making a standard system desirable.

Some of these links, such as the electronic connection between our farm and its urban customers, are benign and non-colonial. It seems to be common sense that if the commerce and information sharing is going to be conducted in cyberspace, we all need access that is roughly equal.

But we are still talking about information sharing and, at best, the occasional communication of knowledge. This is nothing to get all religious about. The hard work of figuring out what is wrong with rural America and how to cure it has gone on and will continue to go on independent of the question of broadband access.

Any real and thoughtful analysis shows that rural problems stem from the fact that food is not valued highly enough. This has been a driver of change in rural America, and not change for the better. It didn't start recently. When the financial powers that be set their eyes on the west and engineered the legal structure with which to steal it from the Indians, they had cheap raw materials on their minds. They settled the lands with hungry and desperate European immigrants, built the railroads to get their production back east in a manner that left them in control of price paid, and did whatever further processing was needed in urban centers such as Chicago — again by use of underpaid immigrants. This stranglehold on midwestern and western farming lands continues to this day, aided and abetted by technology that has come to the farm — always to make it bigger, never better.

The cost/price squeeze that results from farmer control of only production, but never price, has been well documented. What is not so well understood is the impact of cheap food upon the quality of the farming and the quality of life. Suffice it to say that oil is not the only industry destroying the Gulf of Mexico. Agriculture is, too, by means of farming shortcuts. Like all other Americans, we farmers have been coaxed and driven to think of ourselves as economic animals only, and much of what little we knew of good farming practice and culture, along with all of what we could have learned, has fallen by the wayside in our race to keep up financially.

The bright spot in all this lies within a certain small part of the buying public who have decided that food price should be a secondary consideration, and who are willing to buy their food based on how it tastes and how it was produced. This gives a certain few of us a little breathing room in which to begin the building of a new kind of farming economy, one that will be kind to the land, fair to us farmers and straightforward and honest with the eaters of the food. It is a small chance, and there is no guarantee that it will come to fruition. But it is real in a way that rural broadband access as a rural economic savior is not.

ENABLING THE SCOUNDRELS
DEER RIFLES

There are those who think the entire United States Constitution is the second amendment, the one about the right to keep and bear arms. They proudly state that their guns are what secure all of the rest of our rights. And so they strut around carrying them at health care hearings and wherever else they can get in front of a camera. These people are dangerous to anyone who disagrees with them. They may be dangerous to the President if they can ever get within range, and they are lethal for congresswomen and men who would like to meet their constituents. They are deadly for innocent bystanders. But they are no threat to the government of the United States, no matter that the government is interested in trimming down as many of the people's rights as it can.

I believe in the second amendment. I keep guns. I use them to execute the occasional pig that is down on its luck healthwise, and to keep the skunks (the pretty ones, not the two-legged kind often found in Congress) away from the yard when spring meltwater floods their culverts. One of my guns, a Stevens 20-gauge single shot, means a lot to me, as it was a gift from my father. I am not a hunter. I was for a while in my youth, though never a very successful one. Since, I have let farming interfere too much with hunting and fishing and other worthwhile activities. I like hunters. I wish we had more of them. They perform a real service in keeping the game animal population in check and also in injecting a sense of reality into a population otherwise so very divorced from the practicalities of life on earth.

But really — deer rifles and automatic pistols against the most advanced military the world has ever seen? It is past time to admit that the second amendment, along with Jefferson's musing about the tree of liberty needing to be watered occasionally with the blood of patriots and tyrants, is a relic of the eighteenth century. I don't mean that the amendment should be cancelled by a later one, or that it should be the excuse for rewriting the Constitution, but simply that it should be viewed rationally. Perhaps then, big city mayors could have a clear chance at getting some of the guns off their streets, thus helping to keep some of their kids alive.

And it certainly is useful to notice that the people of Egypt and Tunisia have begun major change in their countries with no firearms.

So if the second amendment cannot logically protect and guarantee our liberties, what does? What guarantees our rights of free speech, freedom of press and to assemble, freedom of religion, to face our accusers and obtain fair and speedy trials? Well, who runs the country? Given the events since the Citizens United decision, which gave corporations unlimited ability to fund political ads, only a most fanatical libertarian might argue that the corporate elites are not in charge. Many of us have thought for years that the corporations are not only in charge of the government, but of the entire world. The international structure of agreements on free trade pretty much demonstrates that fact. We who are accustomed to defending our rights against politicians will need to see through the screen to the entity that controls the politicians.

How do we defend our rights against the real enemy of our rights, the corporate elite? The corporations have made themselves more necessary to us than the government ever was. Thus, and so not illogically, many who are always and rightly suspicious of government seem to be able to trust corporations with everything.

We have become absolutely and completely dependent upon them. We are born in a corporate hospital, cared for all our lives by a doctor who is an employee of a corporation, wear corporate-made diapers, are entertained by corporations, eat corporate food, drive corporate-made cars that use corporate-owned petroleum. We are "educated" in an institution that if it is not already corporate, understands its mission as making future corporation employees and, especially, customers. We work for corporations, are thrown away and then nursed in our declining years by them, and are finally cremated or buried by a corporation. Notice sometime how very little of the "final respects" to a deceased person at a funeral are carried out by the representative of our religion or by the deceased's closest friends and relatives, and how much is handled by the corporation that is in charge of proceedings.

How do we make an institution that is that controlling of us respect our rights? Only by making that institution understand that we can live without it. The second amendment cannot secure our freedoms in this atmosphere. Only competence can. Egypt has no second amendment. But Egypt, with thirty percent of its children malnourished, has no shortage of people who know how to live on the edge. We can start our liberation by getting control of our buying habits. Can we get competent fast enough?

ENABLING THE SCOUNDRELS

STUNNED

My sister, who grew up farming every bit as much as I did and who has spent most of her adult life working with farmers in addition to running her own beef grazing business, is amazed at how farmers will take advice gladly from anyone who is selling them something, but only reluctantly and suspiciously from anyone else. It is a fault line, folks, that runs deep into our character. Farmers will ask their fertilizer dealer how much to put on, and they eagerly await each new manufactured seed from Monsanto.

This is costing us, big time. Check John Deere's annual earnings report. Ingersoll-Rand, which makes one of my favorite gadgets, the Bobcat, pretends to be headquartered overseas to avoid having to pay taxes in the U.S., thus ensuring that we all pay more.

Follow the money. We were not necessarily born this way and we didn't learn it at Grandfather's knee. But we were taught, slowly and carefully, by the people who stood to benefit from our prejudices. This facet of farmer imagination was built in the corporate boardroom the same way racism is encouraged there, and for the same reasons, because it suits them to have the working people — including the farmers — divided against each other, making them easier prey. But the neatest trick the corporate puppet masters have been able to pull is to get us into the frame of mind where the farmer across the fence is the enemy, and the guy peddling the easy solutions to everything is our friend.

Once they got the basic idea in place, it easily progressed nearly on its own. After all, the farmer across the fence probably has the idea that crop agriculture on ever-expanding acreages is the only approach, and he may well try renting land away from me.

Agriculture is regularly criticized for letting its soil go down the river, for mistreating animals, for polluting the air, for using up the roadways at harvest time and...the list goes on. When we notice that the ones making these statements always seem to work and sweat less than we do, have occupations that are more secure than ours, and have their health care given to them, it makes it that much easier to succumb to the blandishments of the corporation whispering friendly words in our ear, while it slips its hand into our pockets.

The difficulty is that this misperception is destroying our rural communities. Suspicion of neighbors tears at the community fabric. Just ask someone who lives in a dangerous part of any city. They know. When I perceive my neighbor as my competitor and suppliers as my friends, I effectively throw my loyalty outside of the community that needs it and toward people and institutions that will use it only to make profits from me. Choosing a business-only deal over the possibility of a deep and wide human relationship carries consequences. Thinking of our neighbors as competitors means that we have essentially no place to rest and relax in the world — no group of people who will "look out for our back" or extend a hand when we need it. Lack of a neighborhood or a community is worse than lack of money.

This mental and emotional gap shows up in pretty strange places. Once, several decades ago when all the schools in this area began to have trouble keeping their doors open, we went with the 4-H club to a neighboring town to view a beautiful flower garden. The owner was a woman in her seventies, well known in the entire area for her flowers. After talking for some time about the plants and what she did with the soil to make it produce so dependably, I asked her how she felt about the tax referendum I knew she had upcoming in her school district.

"Oh," she said, "I will vote against that. You see, all my children have moved away, and so all of my grandchildren are in school in the city suburbs. I have no reason to vote 'yes.'"

So I asked her if she knew who would come to pick her up and take her to the hospital if she had a serious health problem at home. She did, and named their names, most of whom I knew. So I asked, knowing the answer, if any of them had children. She said they did. Then I asked if she thought those children needed a school. She looked stunned.

I have since asked myself why on earth any rural person should be stunned to face that particular issue. But it is fact that many of us would be stunned, which is as good a sign as any of how deep the damage runs.

It would be hard at this point to think of a more pressing problem than land transfer for all the farming communities in the country. As the land gets sold to stock market refugees and other urbanites for investment purposes, any possibility of what we have had in agriculture existing into the next generation is fast leaving us. A healthy community would be figuring out ways to cope. But it is difficult to see how this could be done today, given the predisposition we have to competitiveness and individualism.

ENABLING THE SCOUNDRELS

We farmers who graze our livestock, as well as any other farmers who are trying do things a little differently in either production or marketing, need to know better, and act upon that knowledge. For the most part we do know better. An example is the Sustainable Farming Association of Minnesota, which has as part of its philosophy and mission statement words about the wisdom within ourselves and the development of sustainable farming systems through farmer-to-farmer networking. We need to think about the meaning of that, put aside the usual thoughts of hopelessness, and stop listening to the easy modern voice we carry around in our heads that tells us to just go with the flow, enjoy, and not worry.

None of us can fix the problem, it is true. The problem has been a long time in the making and has been made by many people. But all of us can do something. It is up to each of us to figure out what that is.

CONVERSATIONS WITH THE LAND

CHURCH & CAPITALISM

The working class comes in for criticism, much of it justified. It is, for one thing, arrogant in its ignorance. Don't take my word for it. Go into any working class bar, order a beer, and spend a little time listening. After a while you will want to scream that there are libraries, for God's sake, and they are free and public. It is excruciating for anyone in the working class who like me was raised to believe that we are all born ignorant and that it is up to each of us to do something about it. It is as if we think we are entitled to have learned nothing of the world around us or our own history, and yet somehow live like kings.

The working class is militarist. Partly this comes from the fact that the working class fights the nation's wars — wars caused mostly by other people with religious or financial irons in the fire. Adding to this is the fact that increasingly the military provides the only opportunity for a decent living for anyone without extensive education and/or connections in the business world. The working class is by any measure the most overtly patriotic segment of society, which could be thought of as an attempt to make lemonade out of lemons. Certainly it is not difficult to detect in working class patriotic pride the undercurrent of resentment engendered by the feeling that the rest of society owes us for our sacrifices of time, health and sometimes life itself in the nation's wars.

The working class has trouble recognizing itself. It was the working class that sat on rooftops in New Orleans after Katrina, waiting mostly in vain to be rescued. But since that was the black version of it, the white working class was free to indulge itself in its usual fantasies about welfare cheats and freeloaders. It is difficult to estimate how much stronger the working class might be, or how much more effective organized labor, if it were not for this tendency of ours to be led around by our racial hatred and fear.

These traits did not just happen; they have been built in reaction to found circumstances over the course of many centuries. Labor or "working class" as understandable constructs developed out of the peasant class as a necessary support for market capitalism and industrialism beginning more than five-hundred years ago. This development appears to have been aided by the Church which, in order to stave off further deterioration of its dominant position in society, cut a deal with the developing economic elites, giving industry more or less complete control over the people in all things economic in return for their regular attendance at services and the possibility of a few coins for the collection. Thus began the segmentation of the human spirit. The Church still views itself in this way, speaking to every aspect of human life but the economic, with the exception of the Anabaptist sects and the Quakers.

Capitalism succeeded in part by segmenting the human even further, this time mainly into the worker/producer and the consumer. Henry Ford held these two aspects in an uneasy balance when he talked about the need for his workers to be paid well enough to buy his cars. Labor since then is expected to show up, work hard for eight to twelve hours a day for stagnant wages, and then spend like a drunken sailor on weekends while doing a decent job of raising the kids that will be the next docile cogs in the industrial wheel.

But industry has since tipped that balance with the idea of global commerce, which enables it to exploit one group as workers (Mexicans, Chinese) while it uses another (Americans) as consumers. It has not figured out yet how to keep non-working Americans wealthy enough to keep buying, or overworked Chinese docile enough to keep the machinery running. Now industry is busy preaching the sermon that all reality is economic, that humans are merely economic animals with no recourse to any other emotional or spiritual foundation. Not content with merely segmenting us to improve our usefulness, we have now been reduced. The Church, as usual, is mostly silent.

This economic edifice has been built upon the destruction of the human spirit. This is why we have the arrogant ignorance: People are reacting to the despair telling them that nothing they can do will improve their situation and, in particular, that they cannot improve their situation by trying to improve their understanding of it. Huge forces beyond their power to control are in charge, and helplessness is the order of the day. The reaction is empty boasting and the endless feeding of a false and desperate arrogance.

This is the reason for the love of the military. When one has given up on the idea of having any other impact on life, the power and violence associated with the military has appeal. It enables one to be part of a huge force that controls others, even if that control does not involve any particular benefit to those doing the nasty work of imposing it.

CONVERSATIONS WITH THE LAND

And this is why we hate so well and so reflexively. We hate anyone we perceive as getting any kind of advantage over us, because we have learned well over the past centuries that anyone's advantage is always our disadvantage. Separate from each other — even our own families — separate from creation itself, separate from the production of our own hands, we live in an individualist hell. No wonder so many of us are violent. And no wonder the suicide rate is so high.

I am a member of what is properly known as the peasant class. The fact that my family has always farmed, and that I have spent my life doing so, affords me a little protection from the abrasiveness and destructiveness of the spirit life entailed by being a worker in a society that does not value work.

It also gives me the opportunity to be involved in what can only be called the New Agriculture. Some of the fundamentals of this approach to farming and food can be seen as a direct rejection of the straightjacket all of us less-than-wealthy humans have been wearing for centuries. Take, for instance, the linking of farm production with marketing. This is born of an insistence upon control of production and the determination to make the income return to the work of production. Not accidentally, this also makes necessary a cooperation between at least two people, as good producers are rarely good marketers. Considerable thought is being expended in making sure this new relationship does not deteriorate into a situation where one group owns the production of the other. We have had five centuries worth of that, and it is time for a new thought.

ENABLING THE SCOUNDRELS

FAILURE OF CITIZENSHIP

It is impossible not to notice the stranglehold the corporations have on our political system. From the Wall Street bailouts, which are a blank check to the financial system for destroying our economy over the past three or four decades, to the way the insurance companies are dictating the terms of health care "reform," it is evident that the fix is in. Further indications are the killing of the consumer bill-of-rights legislation, the furor over the consumer protection provisions portion of financial regulatory restructuring, and the mangling of climate legislation coming out of the House of Representatives. In fact, the only corporations the government gets tough on are the automakers, who actually do occasionally make something of value.

The people, Americans all, are harmless as a flock of sheep. And so we must wonder if the essential American character has been remade. We are now a long ways from the Nineteenth Century's indomitable Irish immigrant who, after being shipwrecked just offshore, waded to land while yelling at a group of people nearby, "Who is the government here? And whoever it is, I'm against 'em."

So it seems at the least useful to ask if anyone can be dependent on a corporation for every necessity of life — job, food, clothing — as well as electronic entertainment and political opinion, and still be capable of participation in a democracy that seeks to control and use those corporations for the public good. The scope of the decisions left to ordinary citizens has narrowed until most of what we get to decide doesn't amount to much. Or that is true until we start to swim upstream, in which case each decision gets to be vital again.

As we lost huge numbers of farmers over the past seventy-five years, we lost also the ability to think and act independently in the population at large, which I think is vital to the working of democracy. I do not mean to belittle the difficulty of operating any other business, and I do admire anyone who steps out of a job and takes on the responsibility and work of going out on his or her own.

But we operate two business entities here, one of which is a farm. And I have spent considerable time trying to figure out why the farm is so very difficult compared with the business that sells our meat.

What I come up with is that the farm works with living things — with the entire biological system of earth as encountered on one small piece of land — while the business works with inert substances. In the business, a question arises that needs an answer and an effort is made to find that answer, perhaps considering and discarding several before finding the one that fits. Then the question is solved and the problem goes away. Next question?

On the farm, it is as if the question changes while I am looking for the answer. Or maybe my looking for the answer is what changes the question. After all these years of trying to get this right, the farm seems more and more like a capable and persistent brother I am not going to beat, no matter what I try.

Take bloat for example. Since we started grazing here, we have tried everything to deal with our occasional bloat problem. Move them ahead in the afternoon, we have heard, and never in the morning. Make sure the grasses are succulent. Use bloat block. Give them a small break only. Acclimate them first, and then provide a large enough break to last several days. Graze only dark green alfalfa in at least half bloom instead of the light green, fast-growing stuff. Watch them in May and September, as all other months are no problem. Cut ahead, and let the forage lay for 24 hours. And so forth.

Every piece of this thinking has failed us at one time or another. At the same time, all of it seems somewhat true, or at least true at some times and not others. This spring, for instance, my thinking was that part of the problem was that alfalfa (which I like for the use it is able to make of our deep, wet soil profile) was restraining the grass in the sward. So I figured that after one whirlwind rotation through the paddocks we would slow way down and let the grasses mature to help them establish better. The cattle mostly ate the mature grasses, seed heads and all, and we have not had a single case of bloat.

But — and this is an important but — the spring was dry here. I never noticed the alfalfa as being light green. The grasses absolutely raced to head: I thought they were all trying to be bromegrass. So was it management or weather? Maybe it was both. What if it was neither? At any rate, I will know what to try next year.

Or will I? Forty years of farming, the saying goes, and no two alike.

ENABLING THE SCOUNDRELS

It is this dealing with a business that changes right in front of my eyes in ways that I would never expect that is so fascinating to me. It keeps me getting up in the morning and looking around with fresh eyes, and wondering what the day will bring.

But this is no longer ordinary behavior in the modern world. People do not aspire to it. This is evident from the real shortage of people who want to sign up for this kind of work.

When the corporations used the power of the government to drive so many of us out of farming in the seven decades following the end of World War II, they knew they wanted more of us in town dependent on the job to help keep the wages down. They couldn't have known that they were also fostering the conditions for a failure of the citizenship necessary for democracy.

Or could they?

CONVERSATIONS WITH THE LAND

MONEY & MORALITY

Supreme Court Chief Justice John Roberts and his hardy majority of conservative "strict constructionists" just turned loose the corporate dogs by allowing our most powerful institutions unlimited spending in political campaigns. This ruling allows corporations to, for instance, buy up ahead all radio and television time immediately before elections, in the process turning our founding fathers on their heads and their Constitution into something none of them would recognize.

It is pretty telling that Justice Scalia opined that corporations are pretty much indistinguishable from their owners anyway, so freeing them to buy politicians at will is a kind of business as usual. I, being of somewhat more modest means, have not so far been able to purchase a single politician, more is the pity.

The idea that corporations should be treated like persons goes back at least to 1886, when Supreme Court Chief Justice Morrison B. Waite pronounced himself and his entire court entirely satisfied that corporations were indeed people, and that he wished to hear no oral arguments on the matter in the *Santa Clara County v. Southern Pacific Railroad* case. A century earlier, Thomas Jefferson had been alarmed at the power such an idea would give large financial institutions.

So corporations are "persons" who may never get old or be killed in war, but with rights of free speech and the right to sue and be sued. How being sued works out in practice is problematic due to the difficulty in pinpointing who the corporation actually is so that it can be held liable. Generally, courts have settled for fines, often so light that the corporations can pay them out of petty cash.

While the legal right to speak and write about products should not be troublesome beyond its tendency to involve a great deal of lying (thus distinguishing it from politics not at all), there is another generally recognized attribute of corporate "persons" that does carry through all its actions. This is the fact that corporations have been allowed to shed a good deal of the responsibility, or liability, inherent in whatever activity they pursue.

Corporations use this second attribute for capital formation. Mutual

ENABLING THE SCOUNDRELS

funds, where our money goes a-hunting worldwide for the best returns, are one version of this. We as individual investors come in for the profits our money generates, while someone else gets stuck with any bad effects of the activity that generates those profits. The poverty in Haiti is a good example of this. So are the coalfields in Kentucky.

We need to reconsider these aspects of corporate personhood in the light of what is happening to us, and our economy. The fact that our laws enable corporations to essentially shield their wealth source from the consequences of their own bad behavior is directly behind the current decay in the economy and the criminality on Wall Street.

And the fact that we allow corporations to live on forever has to do with our own willingness to give over important economic decisions to them, to the point where they get to be primary and intergenerational drivers of our economic reality. We as a people get to be affected, rather than being in control of them. It is as if we do not really want control, as we have had a childlike faith in the big economy, corporations and all.

Adam Smith, the father of modern capitalism, thought that a system of morality needed to be brought into capitalism from the outside, from the culture and the people, because capitalism itself does not generate morality. We are seeing how right he was with that. When we decided that corporations should be allowed to engage in economic activity pretty much without fear of their individual investors being held liable, we separated money from morality.

We can't do that. Humans aren't made that way. The results are nearly unrestricted bad behavior on the part of corporations — especially in the financial area — enabled by a kind of helplessness in the general population.

Corporations have had a growing dominance in the economy, and thus an increasing influence upon our government for most of its more than two centuries of existence. The spectacle in Washington about bank regulation and health care is sickening to behold for anyone who cares about the country. So the latest gift to corporations from their enablers on the Supreme Court is not new, but really just one more step in securing corporate dominance in both economy and politics.

Farmers who graze their livestock, unlike politicians, are not big into helplessness. Most are former conventional farmers who have thought their way out of that box, and are now accustomed to succeeding by using their brains and determination.

CONVERSATIONS WITH THE LAND

And the grazing movement is closely connected at its smaller end with a growing group of people who have a little land, even if that is just a house in an urban area. These folks don't like the life they have been raised to and don't particularly trust either the suppliers of their necessities or the general circumstance they see around them, often including the corporation for which they work. The difference between a grazier with one hundred milk cows or two hundred beef cows at the bigger end, down to the guy in South Minneapolis with a dozen chickens in his backyard, is one of scale, not of kind. Even though misunderstandings often crop up, we are all going in the same direction in at least one important way.

The choices we make are making us freer because we are becoming more capable. As we become more capable, the corporate structure becomes less necessary to us. I cannot yet imagine a life without corporations. But I do see a few ways of beginning to de-emphasize them, and that is a start.

Turning the cows to grass lessens the need for tractors and machinery, thus also for fuel, crop chemicals, fertilizers, crop insurance, debt, drugs and so forth. Planting a garden, and harvesting and storing part of it, frees us of some of what the grocery store sells, some of our body fat and some of our dependence on the doctor, the drug store and the workout place. Raw milk gets us away from big dairy and makes us get along with a cow. Or if we buy the milk from a neighboring farmer, we must learn what makes for good farming practices and some of what goes into producing good milk. Backyard chickens help us understand the cycling of energy in nature; the way in which pest insects and table scraps can be converted to better eggs than we will ever get in the store.

Notice how these actions push a certain number of corporations off the center stage in our lives. We have started on that first step, which is to need them less.

This is necessary. We probably cannot succeed in imposing responsible behavior on corporations while we are so completely dependent upon them. The habit and attitude of buying less begins to shift the power away from corporations, and back toward human persons.

CONVERSATIONS WITH THE LAND

HOW WE MIGHT FARM

CONVERSATIONS WITH THE LAND

JAKE'S FENCE

I was raised to farming. Most of my "indoctrination" was carried out by my father. He truly believed that he who could farm certainly would, and that others would just have to be satisfied with a lesser lot in life. When I was a little boy, three or four perhaps, the instructions given for locating me were to "find Jake and look down." The barn my parents built to replace the one that blew down on our former farm a few miles down the road still bears my three-year old footprints in the concrete.

Dad is gone. So is Mother now, and I have been amazed at how very intimidating it has been to find myself in the predicament of being the oldest, at least in that my generation is the oldest. The idea that I am or ought to be speaking from experience is confusing. How did I get so far so fast? My confusion is worsened by the knowledge that I do not in any sense "know" what my parents knew, even though all of us are, or were, farmers.

Last year, I put up a perimeter fence. I hired a contractor to do the installation, but it was up to me to pull down what was left of the former fence and prepare the site. The former fence was my father's fence. I have just the faintest recollection of his building it. It was his in more than memory, for the posts spaced one rod apart in a ramrod straight line and the three strands of barbed wire installed "just so" on the post spoke of him. Dad was a reserved and quiet man who could not be bothered to wash a car, tidy up the inside of a truck or worry about a tree branch scratching the paint of a tractor. But his posts were straight, and his wires were accurately spaced and tight.

The posts were bought new, of the "ranger" U-shaped style and about six feet in length so that, when installed, the wires would be about the proper height to control the herd of Hereford cattle he planned to graze each summer and then feed in the winter. It would not be inaccurate, though perhaps a little risky, to say that I have never felt closer to my father than the week I spent tearing down his fence. I spent the week talking to him. I will leave it up to theologians to decide if he could have heard me or not!

My memory is that later there were bloated cattle laying out in the field that fence surrounded, and from then on I cannot remember any cattle grazing on the farm except the little dairy herd in the continuously grazed, wet pasture. When I was about fourteen, the cows were sold in favor of purchasing mid-weight cattle each fall, which were fed through the winter. The land was put to crops, with just a small planting of alfalfa for hay to start the feedlot cattle each fall.

But times change, and so do prospects. My feeling is that my father was right the first time. The crops that actually did pay the bills for some farmers for a few years in the '70s have become a government-sponsored, corporate welfare activity — an idea Dad would have sneered at were he operating a farm today.

If that generation of farmers had not totally bought into the technological revolution that disrupted the biological and economic cycle of their farms in favor of purchased solutions to greatly oversimplified farm production problems, where would our generation of farmers be today?

The idea that cattle could be grazed in summer and grown into readiness for grain finishing could have morphed into a plan to try to finish some of them by grazing annual forage crops late in the second fall of their lives. If that had happened, would turnips and rape (and who knows whatever else?) have gotten the same academic attention and increase in average yield that corn has seen over the past fifty years? And what about corn grazing? What would the corn plant look like today? Or the cattle? What improvements might have been made in pasture grasses? Can you imagine a situation where we were not marketing against a common perception that beef is bad for our health? And how about the thought that university research might have been responsive to farmer needs, because farmers had the money?

Hindsight is twenty-twenty. And it is likely that it will be just as easy to critique the grazing movement in fifty years as it is for me to find fault with the cropping system I see around me. But it is tough to be objective. I look at the black desert that surrounds my dusty, tan and green pastures each November. Some of that soil will come loose in the winter prairie wind and lodge in my dead grass. And I will make my annual joke about sending the neighbors thank-you notes!

CONVERSATIONS WITH THE LAND

PATTERNS

One pattern sometimes holds true in several different venues. That is true now of doctoring and farming, both of which are in a pretty advanced state of decay. Thoughtful people in both of these areas are wondering how long the current practices and ideas can hold up.

Bruce Lipton, in his book *The Biology of Belief*, points out that it is conventionally accepted that 120,000 people die in the U.S. each year from adverse reactions to drugs. This would make prescription drugs the third leading cause of death. However, this count is obsolete, having been made in 2000. A ten-year survey of government statistics completed in 2003 shows that the real number may be closer to 300,000, making drugs the leading cause of death, according to Lipton. You will notice, of course, that this number never got talked about in our recent "health care debate." But it presents the practitioners of conventional medicine with a picture of the dead end they are on, along with the knowledge that change is no longer just desirable, but absolutely necessary.

There are similarities between the practice of medicine and the practice of farming. Farming and medicine are necessary to human health. Both, if well done, are as much art as science. Or perhaps more precisely, they are the practice of art in the use of science. Whether or not the practitioners particularly want it or even like it, both work with biological systems. This means that there is always going to be some of the "make it up as we go along" about both. Medicine heals the body; farming feeds it. Some insist that if farming provides the right kind and quality of food that it, too, heals the body. And both farming and medicine fail with the failure, before death, of human life.

Agriculture, like medicine, is oversimplified. Just as doctors (especially those who work with the elderly) pile one drug on top of another, often hoping to control the bad effects of the first with the second and third, so, too, do practitioners of agriculture pile one technological solution on top of another. If one dose of Roundup won't do it, perhaps two will. Maybe we will also have to "clean up" resistant weeds with other chemicals. Surely we need every crop to be genetically engineered, not just the main ones.

HOW WE MIGHT FARM

Animals will grow by eating this stuff, if they know what's good for them. We will give them bicarbonate of soda or other minor ingredients in the feed to help their poor, damaged stomachs. If animals fed wrong (grains instead of forages) develop the dangerous e-coli, we will vaccinate them. And if bleach on the carcass doesn't fix it, we will irradiate. Who can doubt that we will soon see handy home irradiation machines for sale in the Walmart?

We will grow our vegetables in monocultures in places that can be most "economically competitive." We will ship all over the world from these places. If the dangerous e-coli is found on the vegetables, we will blame it on the wildlife, hire hunters to decimate the wild populations, and build tall fences to keep them out. Maybe we should bleach the spinach.

What we have here, in both areas of the economy, is a perfect storm of homegrown foolishness! We are pushing on a rope. And since there is really next to no chance the government is going to do anything useful in either of these areas, it is going to be up to us to hike up our drawers and be the adults in the room.

Lipton makes the argument that too much of today's medicine is based on an outdated understanding of physics. The body is thought to be matter only; the role of energy is not considered. He says we are just beginning to understand that each cell relates and reacts to its environment, changing as necessary to live in that environment.

For Lipton, the cell's membrane is its "brain," deciding how much of what substances to let in and what to exclude. Through these simple processes, the membrane controls the cell's activities in its environment, and therefore the health and well being of each cell, which is of course critical to the health of all the cells together: the body itself. Consider that a cell's environment consists largely of the nutrients surrounding it — taken from the food we eat, the water we drink and the air we breathe — and you have a picture of the connection with agriculture.

It is both exciting and alarming to follow Lipton as he makes the argument that it is the cells themselves, communicating with their environment, that carry on a goodly share of the changes seen in an organism during its life, and that some of these changes are passed to following generations.

Darwin, according to Lipton, didn't have the whole idea. As an indication of how little we really know of the life of the body and how cells communicate, Lipton tells us of a health-conscious woman who received a heart and lung transplant from a young man killed in a motorcycle accident. She soon developed a craving for chicken nuggets and beer, along with an appreciation for motorcycles.

And so we might ask: In what long-term direction are we being taken by a nutrient stream of biotech-manufactured carbohydrates and proteins, with chemical residues outside the cell membrane of each and every one of our cells?

Agriculture suffers from ignorance of the basic plant-soil relationship. Farmers today understand the soil and its life less than farmers just a generation ago. This is due to over-involvement of energy and technology corporations in the production, marketing and even financing of crops. We in agriculture also discount the importance of energy, often treating our farms as a simple pile of stuff we can force into making a profit for us.

I think we grazing farmers don't fully appreciate the radical difference of our own approach and what that does for our perception of agriculture. Remember that most farmers think of energy as something they pour into their tractors or produce out of their corn. Some farmers are also dimly aware that fertilizers and crop chemicals are basically energy. Few indeed have any idea of the kind of total energy deficit their farms run, or how exposed this makes them in this climate of increasingly expensive and undependable purchased energy supplies.

Purchased energy is dead-end energy. It is used for a single purpose and then disappears from the system, except as waste product and cost overhang. In contrast, nature's energy flow produces no waste. It costs nothing other than having an understanding sufficient to see it, and sometimes the effort to enable it. It operates in a cycle coming up from the soil to nourish plants and then the animals and, upon death or sometimes the reproductive part of the cycle, these enter the earth again, going through a process of regeneration preparatory to emerging as new life once more.

The catalyst for all this is the sun, which is why the entire process is sometimes called the solar chain. It is our understanding of this cycle that gets us to graze our animals whenever possible, to stockpile forage to be grazed in place if we can or, in a somewhat more complicated and less efficient way, to operate a farm sufficiently diversified and well planned that it mimics the natural cycle in growing feed for animals whose manures are returned to enrich the soil.

Biotech seeds, crop chemicals, commercial fertilizers, and animals in feedlots attempt to turn this elegant arrangement into a simple, factory-style line — from production through use — ending not in regeneration, but only waste. Waste of soil, because its fertility is not linked to its own life through regeneration. Waste of the farm's chance to be profitable and satisfying, because it is now merely a factory. And finally, waste of the consumer's faith that the food he purchases is good and healthful. Unhealthful food is often supplemented with too many pills, so we have yet another connection between medicine and agriculture, this one a negative.

The sharing of patterns is an indication that great change is afoot. We can see it not only in agriculture and medicine, but also in government, economy, education and finance. The old forms are dying of their own contradictions, and are doing a great deal of thrashing about in the process.

Meanwhile, something new is trying to be born. It is exciting, but also terrifying. As with modern medicine, any enterprise based so squarely upon such simplistic delusions as those in modern conventional agriculture can only destroy itself. Let us hope we are not all destroyed in the process.

CONVERSATIONS WITH THE LAND

ART & SCIENCE

We have many pounds of carrots from our garden this year due to the nearly steady rains all summer. They are heavy and thick, though misshapen due to the fact that it was too muddy early on to do a proper job of thinning, a task we are reluctant to perform anyway. The bounty has begun to show up in soups here, in the potato and carrot and cheese "hutzput," and in several hot dishes. (Casseroles are called hot dishes here on the Northern Plains.) A hot dish made of our potatoes, our ham, and our carrots is a culinary delight, as a food writer might say.

To me, it tastes like home. The idea that taste may be connected to place is a new discovery to the foodie elites, old knowledge to those of us accustomed to partly taking care of ourselves, and simply invisible to laboratory science. And it is invisible to laboratory science not because it doesn't exist, but because science chooses not to look for it.

As a practical matter, science serves wealth in our system. But farmers have used science for centuries. For most of that time, the science has been close to the farm and personal to the farmer. It has been practical and useful knowledge gained by a family farming a certain piece of land or by a farming community passing knowledge, both down the generations and around the community of farming people. The knowledge was intimate and close, true to the land and the people at hand, and perhaps not as applicable in the next state or even the next county, let alone on the other side of the world.

For instance, my carrots may not grow as well nor taste as good several miles away, let alone on the other side of the Corn Belt. We know this; it is common knowledge. But several centuries ago, into this long-standing practice entered laboratory science and the development of a generalized scientific understanding of the earth, its elements and its creatures.

For instance, in the last century or so, Charles Darwin has caused a great deal of upset and confusion among religious folks, including religious folks who are also farmers. Darwin's account of his voyage on HMS Beagle and his observations on the Galapagos Islands led him to muse about natural selection in the natural world, and thence, after some development of his thinking, to the writing of *The Origin of Species*, in which Darwin traced the theoretical development of the various species over great stretches of time by natural pressures and changes driven by the environment. It is today pretty well accepted in the scientific world.

Darwin was born in Shropshire in England in the middle of the nineteenth century. Shropshire is home to the Shropshire breed of sheep, specifically bred to do well in the lowlands. The local farmers also bred the Clun Forest and Kerry Hill sheep, along with Hereford cattle. All of these breeds were painstakingly adapted to do well in the different geographies within one English county. The farming families that Darwin grew up among did this "creating." And it was mostly done by the time Darwin showed up on the scene.

These farming families, mostly pretty much illiterate, beat Darwin to his own conclusions! They did it by natural selection, but with a difference. The environment, or the farm, had its say in which animals were saved for breeding, but so did the farmer. The breeding was done to fit the animal to the environment, in this case the farm. I wonder if Charles Darwin noticed what the farming families around him had done before setting sail on the Beagle.

Darwin has been modernized, which is to say dumbed down. Most of the assumptions he started with and that went into the Origin of Species could not have been argued against by any serious farmer. But now there is social Darwinism, by which the weak and poor and helpless among us should be allowed to starve and die to keep the species strong. And there is the idea of the secular perfectibility of the human race, which I seriously doubt Darwin, as a lifelong Christian, could have espoused. It is important to notice, though, that both of these perversions of Darwin's thinking spring from the same idea of science that gives farmers such fits. We are tempted to ask which humans should be allowed to die, and when. We wonder what that philosophy does to our humanity. What does perfectibility mean? What is perfect? Isn't that a nonsense word when applied to humanity?

Farmers properly should have no problem with science. They have used it for generations. What else but science is the practical, lifelong learning and the adaptation to reality in which every farmer invests his life? What upsets about science as it is practiced today is its extension into philosophy and belief, along with its refusal to consider that the study of reality is always complex and only reducible to simple, testable principles by doing some violence to its applicability in any given situation. On our farm, for instance, we must use hogs that have been adapted by this general science over the last several farming generations to fit the needs of the packing, processing and marketing industries — not any particular farm or group of farms.

For farmers, the given situation is always the critical thing. The general principle means nothing if it does not work on my farm. And my farm is never a simple question, but always complex. Farming has always been art plus science. At least until science matures, art will need to inform when and when not to implement science, and which science to use.

HOW WE MIGHT FARM
ELEGANT SOLUTION

Good farming is based upon careful observation. Observation is discouraged, if not disallowed, by modern farming, with its fifty-foot wide equipment and track tractors. The operator of one of those would never have noticed the pheasant in the fenceline from which I was pulling thistles, and the fact that she was brooding six eggs yet on the twenty-sixth of July. Certainly the operator of that tractor — who would not have allowed the fence in the first place — would not have thought to wonder if she would return to the nest from the oats field or, if she did, whether she would be able to tuck in the one egg that had been displaced by her hasty departure.

If we are to have a good agriculture, it will have to be a "post-modern" one, whatever that means. And we will have to go back in our minds to what came before to see if there is something we left there that we need in this new age.

Such as the chicken. Chickens had a place on all farms of the 1950s, including both the excellent ones and the really lousy ones. That small flocks of chickens are a part of farming's past does not by itself recommend them. But their impact on a farm's insect life certainly does.

Our market hogs spend the last four months of their lives in hoop houses that have self-feeders at the south ends. Hogs being hogs, at least some feed is wasted and mixed with manure, ending up being pushed under the gates where it can form a damp little windrow in the sun. This is a first-rate fly hatchery. But our ranging hens don't let that windrow alone from one day to the next, and that has a very positive impact on the fly problem on the feed yard. Our fly situation is worse for the cattle in the pasture than on the yard where we feed hogs, a fact almost impossible for most farmers to believe. I can see the benefits of chickens following cattle on pasture.

It is important to note that our chickens are the more durable Rhode Island Red and Barred Rock rather than the Leghorn. Leghorns are more profitable in a conventional system, where they return more eggs in a shorter life for the feed fed. But they are not as able to "get their own," which is the most important attribute for anyone expecting chickens to be useful as well as productive.

Observation is critical to economical farming operations. In the case of the chickens, the potential expense of buying insecticides and the risk of using them has been turned into a small profit opportunity.

There are other things I have noticed in the ten or twelve years since awakening from my conventional agriculture sleeping sickness. Bromegrass and quackgrass are similar, but quack seems to like disturbed soil. We have always had a big problem with quack as a weed in our tilled rotations. Meanwhile, the little nooks and crannies around the yard and buildings where we cannot till, and may or may not have mowed for weed control, are always bromegrass. The road ditches in our area are brome, too, with a little reed canarygrass in the lower parts. Sometimes there will be a small colony of quack or annual broadleaves where the pocket gophers have been busy.

Thistles like disturbed soil as well. We can see this because we have a healthy thistle patch wherever the hogs have made the pasture bare, such as around watering points and in heavy traffic areas. I also made the mistake of continuing the winter pasture-feeding scheme for the heifers for several weeks too long last spring. The foot traffic in the thawed mud turned a new seeding of reed canary into a stand of thistles, with the grass underneath.

Observation tells us when the hogs tend to get sick with what, when the cattle need more attention, and that we don't seem to have any natural predators of the pocket gopher here except for one lean, orange tomcat that works pretty hard.

Observation should be a normal human activity, but for too many of us in this televised and computerized age it is not. When we farmers notice our farms and what is happening on them, including what responds to what, we are in the company of bird watchers and poets. When we act on the knowledge that comes from careful observation, we join the company of "pre-modern" farmers, many of whom are farming today in primitive cultures.

Modern farming with its technical solution to every problem is what got us to where we are. But that technical solution seems always to create at least several more problems, one of which will be that the money is not readily available to pay for the solution.

A solution based in our knowing where we are, what we are dealing with and even who we are will be what I have come to think of as an elegant solution. It is elegant because it fits, and because it solves several problems or creates several opportunities at once. Chickens around the hog-feeding hoops is one of those solutions.

HOW WE MIGHT FARM
FUNDAMENTALISM

Today I listened to a young journalist talk about her quest to find the causes of violence and genocide in the global South. She had spent time studying the matter as she practiced her trade in places from the Sudan and Darfur to Indonesia, testing what she learned against her theory that religious fundamentalism, both Christian and Muslim, was the cause of most of the mayhem. She found that yes it was, at least sometimes and in part.

The journalist also found that it was not always the cause. A fundamentalist believes that he has the only true understanding of and connection with God, and therefore believes everyone else to be wrong on this most critical issue. But she said that it does not follow that a fundamentalist will murder to further that belief and agenda. Many factors are involved, not the least of which is the relative security of people's lives, their economic security including availability of food, and the violent games being played out all around over ownership of resources by the world's elites.

So this journalist discovered particularity in her very attempt to understand the world in general terms. When trying to predict what a fundamentalist will do, it matters who the fundamentalist is and what his circumstances are.

Similarly, when asking what a good, profitable, ecologically sound farming practice is, it matters which farm it will be applied upon, and which farmer will do the management and work. Our agricultural intelligentsia has us several generations down the road toward one size fits all, which results in great wealth to the makers of the tools, and great anxiety for the farmer trying to fit the general (and extremely expensive) tool to the particular need at hand.

We have the draining of wet spots and the irrigation of veins of gravel. We level hills so that our machinery guidance systems will work better, and we begin to petition township boards to abandon sections of country road so that our big tractors won't waste so much fuel turning around. We clump up the hogs and poultry in huge buildings to protect them from "environmental stress," and then medicate them to protect against the microbe hot house we created. Roundup Ready biotechnology leads inevitably to weeds we can no longer control.

What is in short supply is a willingness to say that the answers to all agricultural problems depend upon the particular farm, and to trust a farmer to find those answers.

A sustainable agriculture class from a nearby college came here last week for the afternoon. It was a beautiful day. We spent four hours touring the farm's livestock operations and talking about agriculture and our farm in particular. I was struck by how often their comments and questions led me immediately from generalities into the particulars of this one farm and its businesses and its people:

"Why do you pasture the gestating sows?"

"Because they need the exercise. And the extra space keeps them in a better mood. Because the cattle benefit from the sows in that they do not bloat as easily with the sows eating the legumes. The sows benefit from a large amount of fiber in the diet; when they farrow, the resulting large digestive system makes them able to eat a large quantity of concentrated grain ration, and this helps them milk better for their piglets. We have had many fewer problems with mastitis and sickness after farrowing. And the pasture benefits from a better and more complete grazing than the cows could do by themselves. The sows help keep the flies down on the cattle in summer because of the way they spread the manure pats out and dry them out."

"So then, you have the gestating sows on pasture, but you didn't farrow on pasture?"

"Yes, we didn't farrow the sows on pasture this year because it rained all summer and we felt they needed to stay on concrete to manage the farrowing properly. And we have been thinking about moving away from pasture farrowing because our meat sales demand a year-around supply and it is extra work to set up and take down the pasture farrowing in return for just two out of our seven farrowings per year. Also, our farm is poorly drained and low in most of the areas adjacent to the barns, and pasture farrowing in those conditions takes extra management."

"Why do you free-range hens on the yard?"

"To keep the flies down around the feeding hoops. And to turn wasted feed and grain into a good product. And because we like the taste of the eggs."

Fundamentalism comes in many more versions than the religious, of course. There is market fundamentalism, the belief that profits are god. And there is technological fundamentalism, the belief that machines are god. Belief

HOW WE MIGHT FARM

in the necessary goodness of growth in market systems is another fundamentalism, as is the belief that the failure of some always adds up to benefits for all.

Because it is subscribed to by very powerful people and institutions, agricultural fundamentalism causes much damage to rural people and communities. Land is farmed poorly with total disregard for its health, and some land is not farmed at all — a remarkable circumstance in a country with hungry people living on the streets.

General knowledge is easily available, and it is therefore a part of wisdom to know when to turn from it to look more closely at any given situation. Understanding a thing, especially a biological thing, in its particulars and in all its complexity is hard-earned knowledge. My visitors last week seemed interested in learning to do that. Therein lies cause for hope. Perhaps we can yet grow another crop of farmers.

CONVERSATIONS WITH THE LAND

DURABILITY

One of the truly frustrating aspects of the recent conversation about climate issues in official agriculture circles is the way in which grazing systems get overlooked in any talk about carbon sequestration. It's all about chemically supported, no-till systems of crop production. This shouldn't be surprising, I suppose, since one of the real successes of the grazing movement so far has been its ability to maintain a steady focus on the concept of lowering production costs, thus creating no great profit opportunity for any of the corporations that are pretty much in charge of our agricultural thinking.

Consider the progress made in grazing so far. Since its rebirth in the United States following the appearance of Alan Savory and his concepts of Holistic Management twenty or so years ago, graziers have moved from talking about planting nothing but posts the first few years, to considering the grasses and legumes commonly available, and then looking abroad for a few more exotic forages to plant. This was accompanied by a relatively short love affair with all things New Zealand that lasted until many of us began to figure out that New Zealand's maritime climate and regular and steady rainfall meant that many of the practices there could not be transferred wholesale to the severe continental climate of our American Midwest.

We began to talk about swards of grass, forbs and legumes, and discuss the durability and adaptability of various mixes to various farms and to milking cows, or beef cattle, or sheep, or stockers, or — horrors! — even hogs. We started to focus on what we needed to get through the year, or at least the grazing season. Annual crops were considered again for grazing, and achieved a regular place on some of our farms to fill in the seasonal growth flat spots of perennial plants. We began to modify some of the first fences. Those of us who milk cows built lanes and good, user-friendly milking parlors. We got water to the pastures. We started to persecute the universities about the lack of good, up-to-date forage development and research.

We figured out ways to use the buildings left over from the confinement days. As these wore out or were outgrown, we thought of hoops for winter protection. Some of us went to bale windbreaks for winter "housing." Trees came into real importance again here on the prairie, and many windbreaks and rows of bushes were planted on our grazing farms as we noticed that animals don't like wind so much. Many farms began feeding hay on the pastures, figuring that it was best to drop manure where it was needed, rather than be stuck with moving it to the fields all spring.

Then we turned our attention back to the grazing season, and worked on extending it well beyond the growing season. Talk turned to stockpiling grass for after-season use and gaining access to the neighbor's cornstalks for late-season grazing. Rape came into the discussion, as did turnips. Work was done on many farms with turnips or rape, small grains, and field peas for October and November grazing. A few farms are making these seedings work in some years by putting them after the harvest of the spring-seeded grains on the same ground.

Some northern farmers with stock cows started looking to Canadian graziers and their idea of seeding small grains in midsummer and then timing their cutting so that the swaths were soon covered with snow, keeping them in good condition for the cows to dig out all winter. The latest discussion is about height at grazing, grazing periods and timing of return during the growing season itself, with more of us becoming aware that the answers to these questions are going to depend upon our individual animal enterprises and goals.

We clip pastures, or some of us do. But most of us are talking about how many weeds we can live with and what livestock species like to eat which plants. When we get sick of sitting on the mowing machine, we tend to think about donkeys and llamas and goats. We have been listening to a woman who claims to be able to teach cows to eat thistles. We have noticed that if we rotate a little haymaking across the pastures, we can achieve a good bit of noxious weed control by making the hay. If we feed the hay later in the grazing season or in winter right where we made it, we have not hurt the fertility of the ground.

Folks, none of this gladdens the heart of the salesman. Like most graziers I am interested in a tractor with front-wheel assist and a good cab for winter feeding, but no machine salesman is going to winter in Texas on what he sends to my farm on the implement flatbed. Compare that to the no-till row cropper. He has a twenty-four row planter laden with fertilizer and chemical tanks, and residue-cutters being pulled by a two-hundred horsepower tractor that spends the other eleven months in the shed. Then think of the combine to harvest, the trucks to haul the crop to town, the chemicals and fertilizer for the tanks, and the seed bill, complete with technology fees. This is a no-brainer, really. The folks who make their living by farming the farmer are hoping our idea doesn't catch on. We don't offer enough return to their efforts.

So we are probably going to have to get used to being passed over in the carbon sequestration conversation. That is all right, though, because that game is an illusion.

What is not an illusion is that, whether or not we often think of it, we are getting ready for a very real future. That future is about who wins when the commodity markets have fractured due to customer distrust, when the fuel is short in supply due either to a real situation on the ground or banker games, when transportation doesn't work well and steel is too high-priced to use carelessly, when labor is more available than credit, and when our bankrupt government cannot make the handout check good.

Grazing farms sequester more carbon than any others. But that is never going to be the important question. Durability is.

HOW WE MIGHT FARM

FARMING

Annual crops, which are overemphasized in today's agriculture, can nevertheless be made more profitable by a close association with livestock. Livestock will, for one thing, allow a considerable spread in the too-narrow crop rotation practiced today. It is a real luxury not to have to worry about whether the elevator is buying a different or alternative crop when the market is the livestock. This is particularly true of pigs, which can eat nearly anything. But any kind of livestock is useful in this regard.

I have raised a few eyebrows among city friends and acquaintances by suggesting that they need a household pig instead of the variety of dogs and cats they keep. By the time I get so far as to close the loop by suggesting that the pig that has eaten the household garbage should then be eaten by the household, they are generally done with excusing me for my increasing eccentricity. The best I can hope for is a sudden change of subject.

This homely economy is precisely the function livestock should serve on an ecologic and economic farm. The livestock need to use what is not directly useful to us, and then provide us with something of value. The useless items may be grass, clovers, alfalfa, corn, cornstalks, small grains of all kinds, amaranth, beans, peas, gone-to-waste Halloween pumpkins, volunteer regrowth and residues of all kinds. You think of it, livestock can use it. Even feedlot steer manure is better stuff once it has been through a pig.

To see these connections and think them through to any kind of useful action requires a different kind of mind than the kind generally in evidence today on the farms or elsewhere. For one thing, the mind must be able to take a wide view, backing up from the close-up focus of the specialist that sees corn (or beans, or hogs) as the business. This person calculates only those costs accruing to the farm business, while discounting those footed by the natural and human community. Any deficit in the margin of profit is fixed by seeking that margin on more and more units.

The mind that can take the wide view has opened itself to the past, as well as far into the future. It understands the farm as part of the surrounding ecology, economy and culture. It attempts to count all costs, because it knows that those passed off today (such as soil erosion) have a way of showing up in the next generation as bills overdue (decreased productivity).

The wide view is one that attempts to look back and learn what nature was doing here when we came, and then thinks of ways to get the farm to mimic those things on the sound theory that nature has deep pockets, and it is probably best to avoid doing battle with her if possible.

This is known as consulting the genius of the place. For instance, if I have learned that grass and both large and small grazing animals were part of the geography before my ancestors and even American natives came here, then I can, if I am smart, try to configure my farm as much as possible along those lines. The reason I would want to do this is that if nature was doing it before, I can be assured it is going to be easier for me to do than certain other things that have to do with tearing up the soil and planting row crops all the time. Nature will help me do the one, and will fight against my doing the other.

A grassland system is never just plants and never just animals. That would be ludicrous. The plants and animals need each other. They must live together in a kind of harmony or interlocking dependence. That is the blueprint for a grassland farm.

A grassland is not simple either. Its strength comes from the fact that it is not simple, that it has layer upon layer of links and bonds among all its various species so that if one fails because of some unforeseen change, the whole system does not crash, but simply adjusts and goes on. This, then, is the economy of a grassland farm.

Now as a matter of fact, these things are difficult to work out in a farm setting. Seven or eight species in the cropping rotation are a lot to manage, and two livestock species are more than twice the management burden of one. Four or five are unimaginably difficult. Why is this so? Because we are beholden to an economic theory that counts only short-term profit and a management scheme holding that individual farmers must make all farm decisions, and that there should be as few of them as possible.

It should go without saying that we need more people in agriculture instead of fewer. It is astounding that we still have a certain group that greets every wave of farmer exodus with the blessing, "We didn't need them anyway."

We did need them. And we needed them farming better than they ever did, as well as better than we ever have. We need to get over the idea that ag-

HOW WE MIGHT FARM

riculture is a done deal. It is only then that we will admit, even to ourselves, how much we have to learn.

It seems evident when we compare the biological diversity of natural systems to the biological simplicity of human ones that we have a long way to go. And if we start by putting livestock back into their rightful place on the land and at the center of the farm's biological system, we are immediately faced with the necessity of having farmers who understand those kinds of systems, and in sufficient numbers to do the management.

And not only that, the farmers doing the managing are going to need to cooperate with each other as they never have before, so that they can gain access to the beneficial effects that several species have when managed together. This is one of those whole-is-greater-than-the-sum-of-its-parts deals. Anyone who has managed sheep and cattle together on the same pasture, or used hogs in a cattle feedlot, or kept hens around livestock facilities instead of making a facility for them, knows what I mean. This is where the profit needs to be sought, and we will find it is real profit that is lasting because it does not use up its sources.

Seeing the farm as a whole, biological system is the center of a set of social connections that play out into the farm's family, then the local community, and finally to the region, nation and world. This implies an economic system that operates in like manner.

Patricia Love, in her graduate work at the University of Minnesota, made this point in very straightforward economic terms. Her work about changes in dairy farming around Green Isle, Minnesota, pointed out the negative effects on the community as older dairy farmers sold the cows and rented the land to croppers, or became cash-croppers themselves. "Even if purchased locally," Love wrote in her conclusion, "a cash cropper's most costly inputs do little for the local economy."

By now this has been studied enough so that it can be accepted as a given that what worked was the older idea of a diversified farm surrounded by a diverse community providing services for the farm and its family. There is no very good reason to throw it away.

But the fact is that we live in an urban and increasingly cyber economy. And if we look carefully at Love's older dairy farmers, we can see the missing piece of the puzzle of American agriculture. These farms are ceasing because there is not enough profit in them and too much work. The younger generation is not standing in line to take them over.

The answer to the problem is becoming obvious: Farmers must take control of the markets to bring enough income home to properly support the farm. If the income is there, the young will be as well. The problem is our commodity market focus.

In most farm discussions the idea of linking with community never gets taken beyond the little town down the road. But if we do broaden it to include the entire surrounding community, we begin to open ourselves to the change that can happen when people have a say in the food they buy. It is this part of the channel that has never worked in American agriculture, and the smaller, diversified farmers are at the controls of this major tool.

For example, the pork industry has been pretending to respond to consumer demand for several decades. What this amounts to is a standardization of the hog so that the packer can be more profitable. Other than that, although the consumer is generally thought to be interested in "lean," (s)he is never consulted.

In contrast, our family's food business regularly is approached by folks who have quit eating pork because they don't like something about the industry. When we tell them how we raise hogs, they jump at the chance to buy. How many people does this amount to? I don't know. Whatever the number, we will be working to make it bigger.

Small, diversified farms are the only ones capable of this kind of change. From these kinds of farms will spring the kinds of agriculture with enough imagination to once again combine animals, plants and land in intricate systems that mimic nature, and to begin the process of reintegrating humanity into nature's system.

CONVERSATIONS WITH THE LAND

HOW WE MIGHT LIVE

CONVERSATIONS WITH THE LAND

STICKERS & DRIFTERS

The telemarketer who was trying to convince me that I could "earn" a ninety percent return to a play on the stock market was surprised to hear that I didn't deal with criminals. He was so surprised to hear this that he hung on long enough to hear me say that I made my living by working for it, rather than trying to cheat someone else out of it.

Telemarketers, who pop up about three per day on our two phone lines, are closely related to the mosquito. Being a human invention, they may in fact be worse. If we think in Christian terms, we have to reckon with the knowledge that the mosquito is part of God's creation, and that therefore He is well pleased with it, even if we are not.

The incessant yapping on radio and television these days about the new-found strength in the stock market comes from the same impulse as the telemarketer's pitch. It is the voice of the con man and the bunko artist telling us that all is well if the nation's financial system is well.

We know better. Working men and women have found themselves on a slow and steady decline since the early '70s. They have stayed afloat only by the fact that for two generations we have worked more hours, put more people to work at more jobs, and then finally borrowed against our own hard-won equity to keep spending or, in some cases, merely to keep living. We also know that whatever wellness there is in the stock market is due to its raid on the federal Treasury. Obama is just as wrong in this as was Bush.

But the bigger question is about the tendency to equate wellbeing with plenty of money. Reservations about the goodness of large amounts of cash are not being talked about at the head of either party, or anywhere in Washington, or on Wall Street, or anywhere within the confines of the communications empires that control what we see and hear. But it is a discussion familiar at the edges of the power structure. It is part of the conversation in conservative Christian circles, as with the Libertarians and the Constitutionalists and the Progressives. It is talked about among the various "dropouts" from education, from "health care," from the corporate work world. It comes of an understanding of the difference between "stickers" and "drifters," or, as the author Wallace Stegner had it a generation ago, between "nesters" and "boomers."

HOW WE MIGHT LIVE

It has been pointed out that as easy as it is to blame politicians by saying that government regulation of Wall Street failed in the past several decades, this is not useful. The ever-present boomer mentality in the population would have sacrificed any politician who put (or kept) a regulation between a constituent and his chance to get rich. The battle between the boomer mentality, which lives to get rich quick and without too much work, and nester thinking, which wants to build a home, a family and a community, rages in all of us. It is our American DNA. Some of us have more of one tendency, some more of the other. We have a unique opportunity now, as the boomer-made disaster we see all around us slowly begins to heal, to think some new thoughts — thoughts that can be built into new and hopefully better social structures. We can, as the trendy types might say, channel our inner nester.

We farmers who graze our livestock are a small group admirably equipped to participate in that. Graziers are farmers who have separated themselves from standard farmer practice. That takes courage. We graziers have tended to use fewer and simpler machines, to spend more time with our farms and at home, and to make our livings by thinking. We have thought more about nature's involvement with our farms, about the next farming generation, about the ethics of livestock production.

Some of us deliberately limit the sizes of our operations, while others are pretty focused upon growth. But those who do focus upon growth generally seem to hold it as good within a context. For instance, most will say they want to grow their farms or start other farms to include other family members in the business, or to help maintain owner-operated agriculture into the future while benefiting the communities involved. I have yet to meet a grazier who wants to grow his farm into the kind of thing that can be bundled and sold on the stock exchange to provide a retirement on Easy Street. Any graziers interested in Easy Street want the satisfaction of knowing they are there because they earned it.

Is it possible for us American boomers to move away from the ideology of growth as a cancer cell? Graziers might be pointing the way. What might happen if we began to think that boomer values are all right in small doses, but only so long as they are circumscribed and kept under control by nester values? Think of some of the formulations:

- Growth in the business must go hand-in-hand with improvements in careful use of the land.
- Farming success means more time for play and family.

- Increasing the size of the farm must be better for the community. Starting another one might be better by this standard.
- Farms cannot succeed in the failure of their communities and school districts.
- Successful farms strengthen their families.

These points are not found much in farming magazines or in farm management bulletins. They will not appear in the pages of the Wall Street Journal. These practices will not gain us fame and notoriety. But they seem right to me.

HOW WE MIGHT LIVE

RIVER

"River of pollution flows through heart of state" says the lead headline in the Minneapolis Star Tribune. Further down, we are informed that the story will detail how farms and residents along the river's banks have spoiled "what was once one of the most beautiful prairie rivers in the nation."

Immediate reaction in the country would be one of anger. Mine certainly was. I am one of those rural residents operating one of those farms, and I hate having guilt laid upon me. Operating in today's agricultural economy gives me all the pain and hardship I need.

But the article is well done and it is hard hitting. The river is indeed in trouble, and we are responsible. We have been wrong about some things. To plead some extreme view of property rights to justify dumping "surplus" water and soil on downstream neighbors is simply wrong. And it is perverse to argue that a need for "profit" in our rigged agricultural economy justifies more corn piled on our main streets and ever murkier water in the drainage system. We are killing the goose that laid the golden egg.

The picture of green, phosphorus-laden water in a straight drainage ditch, along with the one of small children playing in what used to be a small river before million of gallons of hog manure ran through it on the way to the "Big Ditch," are enough to make any farmer wince. We are responsible; there is no way around it. But is that all the farther it goes?

In reading the lead editorial in that edition, we might think so. After totting up the things gone wrong, such as off-the-chart bacterial counts, the green and murky water and a hefty estimate of Minnesota's contribution to the dead zone problem in the Gulf of Mexico, the newspaper goes into a familiar argument. The farmers and rural residents who affect the river should no more be let off the hook for their actions than the factories were a generation ago, the paper argues. And the cure? Better enforcement. Financial help from the legislature is suggested for the costliest problems.

I would like to suggest that we have an opportunity to take a better path than we have taken before, one that can be followed while not letting up on enforcement of current regulations. That route is hinted at by the last line of the editorial that speaks of public resolve to restore the river to health.

We labor under the illusion, fostered in us by our benighted view of economics, that the only proper role for a citizen wanting change is a political one. Thus we endlessly expect our politicians to miraculously deliver us from the effects of our own, poorly lived lives. What impact would a changed approach to daily life have on the Minnesota River? Certainly changed farming practices might help. But what drives that kind of change?

What would happen if just twenty percent of Minnesotans decided they were going to help restore the river with their whole lives, rather than just their political influence? Imagine the impact if that large a group started looking to buy pasture-raised beef, pork and chicken. The impact would be enormous if twenty percent of Minnesota's food buyers told their grocery stores that they expected to find beef, lamb and milk that was not grain fed, pork that was raised outdoors and eggs in season featuring the bright orange yolk that happens when a chicken eats what a chicken is supposed to eat. And if those buyers also demanded that the food be "Minnesota Grown," this would be economic development of the highest order for rural Minnesota.

There is a great deal of satisfaction to be had from the attempt to live and work in ways that promote the kind of environment and community we want. And there are farmers who want to line the Minnesota River drainage with well-managed pastures featuring tight sod bases that would keep the phosphorus, soil and manure on the land where it is needed, instead of in the river where it is not.

These farmers need a market for the livestock produced in that way. These farmers need the first twenty percent of customers, and then each year more people who will go out of their way to purchase this kind of food while placing steady pressure on grocery stores to begin supplying it. Together, we can make the river run clean. We need merely to act upon our convictions.

HOW WE MIGHT LIVE
READY FOR CHANGE

Consider these words from Time magazine's August 31, 2009 article, "The Real Cost of Cheap Food":

"A transition to more sustainable, smaller-scale production methods… would require far more farm workers than we have today. With unemployment approaching double digits — and things especially grim in impoverished rural areas that have seen populations collapse over the past several decades — that's hardly a bad thing."

Pause for a moment and let that sink in. Time is a center-right publication; it is in no sense a left-wing journal. Did you ever think you would read words like that in the conventional press? Surely the statement itself sounds pretty commonsensical to those of us in the country, but that is because we are rural people and have actually been thinking these thoughts for some time. Time magazine is not rural. Bryan Walsh, the journalist who wrote this, is an urban, upwardly mobile professional type. Urban folks supposedly have had the advantages offered by the takeover of agriculture by technology and high finance.

These words could have been more startling only if they had come from the mouth of an agricultural economist. Then they would have been the equivalent of swearing in church, for in the world of economics you never, ever say that more people involved in productive work is a good thing. Remember that the favorite construct of the economist is that bit about how many people each farmer feeds. This number goes up every year, while the farmers actually doing the feeding are left to wonder where all the people went.

It is not only that agricultural economists condemn rural people to this kind of treadmill toward disaster, but that as a whole they cheerlead for a future where nobody works in making anything, and everybody is instead a customer or consumer. That is why labor unions are so hard to get along with these days: they are made up of ordinary people smart enough to see the future being planned, and know that it can't work.

The whole economics profession, on the other hand, is delusional. And of course conventional agriculture, which has benefited big time from that delusion, is livid with anger over the appearance of the Time article.

CONVERSATIONS WITH THE LAND

Maybe there is hope if a piece like this can appear in the conventional press. The paragraph quoted ends with this from Nicolette Hahn-Niman, attorney and wife of Bill Niman, founder of Niman Ranch: "We are hurting for job creation and industrial farming has pushed people off the farm. We need to make farming real employment, because if you do it right, it's enjoyable work." The truth, of course, is that no one needs to make farming real employment, for it already is. It is just not properly paid.

It would help, as we wrestle with these things, if Washington, D.C., were not full of bought-and-paid-for hacks; if instead there were some real statesmen and women there. Surely you have noticed, for example, that we have been coping with our current economic calamity (real unemployment at seventeen to eighteen percent is a calamity in anyone's calculations, and dangerous to boot) mainly by writing out big checks to Wall Street criminal types. We "fixed" the problems in the auto industry by guaranteeing profitability for GM and Chrysler, who promptly responded by moving more factories to Mexico.

Evidently nobody in Washington ever thought that we should take the same money and instead employ people discarded by the auto industry in various jobs fixing American infrastructure, while letting the companies sink or swim as the case might be. That would have brought unemployment down and infused money into the real economy of food, utilities, clothes and so forth.

This will not happen, of course. But the hope is that we are no longer quite as much on our own. There are other folks dissatisfied with agriculture as it is, and thinking about what needs to happen. Michael Pollan springs to mind with *The Omnivore's Dilemma*, as does Eric Schlosser and *Fast Food Nation*, as well as the documentary movie Food, Inc.

The Time article is short on practical, workable ideas for getting there from here, as things like this always are. But it recognizes the problem, which is a start. If the people who buy food start to see the shortcuts that go into making what should be good food into something that is dangerous to the health of those who eat it — and for farmers, communities and the earth itself — there will be a few, and then a few more, who will help us make some changes. Our farm and business are very much based upon the sorts of people who notice such things and are willing to go some distance out of their way to help make change happen. For that we are grateful. May their numbers grow.

Much of the hope centers upon we who are involved in alternative means of production and the growth of local food systems. It is up to us to make sure that we include consideration of the need for more human hands on our farms and in our businesses in every capital decision we make. Should it be the new building, new feeding system and bigger truck, or can another full- or part-time worker achieve some of the same ends?

Employment is a serious matter, and not to be taken on lightly or to achieve a political goal. But we need to think again that perhaps what we need now more than ever is the neighbor discarded so easily and lightly over the last several generations, either as an employee or perhaps as a contract worker or partner.

It is certain that we need our own young people with us, and the times have never been less optimistic for wage work off the farm. The young need to be treated right, and then have expectations placed on them.

Nothing changes if people are not willing to consider change. The Time article is a good sign that real change is possible. Human minds, both in the cities and in the rural areas, are ready to accept it.

CONVERSATIONS WITH THE LAND

FOOD & HEALTH

We are what we eat. But in another way, we are what we think.

And what we think is that we are consumers. Corporations call us consumers, and we allow them to get away with it. In the same way, drug pushers have been advertising their wares on television by telling the viewer, "Ask your prescriber about our drug." Have doctors no dignity? After many years of formal education and a place of honor in American society, it has evidently not occurred to many medical folks to give that over-dressed drug cartel shill the bum's rush out the door, instructing him or her never to return.

The ideal consumer is a feedlot steer. The steer eats great quantities of what it should not, such as corn and antibiotics. The steer defecates while it eats to make room for more of the same. It is headed for liver and kidney failure, but the plan of the feedlot operator is to get that steer turned into meat before it ruins itself. As meat it may well ruin the eaters — those who think of themselves as consumers. This is not part of the health care debate.

Nothing is more critical to good health than good food. There are things we can do to move ourselves in that direction.

- Learn to cook. Cooking is a necessary adult human skill. Teach your daughters and your sons. This will get you away from the "convenience food" that the food industry thinks of as "profit."
- Get away from fast food. Get it down to once a week, then once a month. From there it is easy to get down to several times a year, and then to pretty much not at all. My wife and I are pretty much away from it and find it doesn't even taste very good to us anymore. Because of how the industry buys, fast-food sponsors an awful lot of human and animal misery.
- Be suspicious of a restaurant that smothers the meat in sugary sauce and puts it on a plate peppered with small chunks of salt. These people are playing on your two major tastes, and it probably means that the quality of the food is not good.
- Examine labels in the store. Don't buy anything with high-fructose corn syrup in the first several ingredients, as it is bad for your health. The food industry is making a feedlot steer of you.

- Garden. This will teach you what really goes into food, and you will eat better. Pressure your community to provide space for gardens for any who want them, whether or not they own any property. It is unconscionable that most of our small communities have a government-sponsored community center where relatively well-off seniors may celebrate their eightieth birthday, but no space in which a poor mother may raise a few potato and green bean plants.
- Buy from farmers markets, and can or freeze what you cannot eat. Make a community or family project of it. You will eat better in the winter, and you will know what you are eating because you can question the farmer you bought it from.
- Get some chickens in the backyard. Let them out on the grass. They will teach you about the real human cost of agricultural production and the silliness of some of your neighbors. The eggs will be wonderful, full of Omega-3 fatty acids and much healthier than any you can buy in the store. Watching chickens is great entertainment, and you can see for yourself that you are not abusing an animal by putting six hens into a small cage to produce eggs.
- Buy meat from a farmer you know. This is the best assurance of quality meats that are grass-fed and raised outdoors as much as possible. Insist that you want your meat processed by shops that pay a good wage to the help and handle the slaughter in a humane fashion. We do not do well unless we all do well. Be prepared to pay up, as cheap meat means human and animal misery.
- Find another recreational activity to replace shopping. Buying and ownership does not give meaning to a human life. Invest the money saved in better food.

To change our health we will have to change many things. Business as usual cannot be allowed to continue over at the USDA, as one farm bill after another rewards the wrong kind of farming. Our universities need to give up their singular focus on pushing as many farmers out of farming as possible, and instead begin to research farmer-friendly systems. Our meat inspectors must be allowed to shut down the assembly lines when they see something bad. Markets and processing systems must be opened up. Anti-monopoly laws need to be enforced.

The list can go on and on. But nothing is as important as citizens — rather than consumers — beginning to rediscover the satisfactions of exercising some control over their own lives.

CONVERSATIONS WITH THE LAND

REAL CHANGE

It is difficult to see how we can keep corporate power out of our elections without real change in every other aspect of our lives. We live in a corporate-sponsored, cradle-to-grave cocoon. Corporate purchase of elections is the least of it.

Part of the trouble for the Democrats and Obama comes from the widespread feeling in the working class that the hard work of real change has not even been approached yet. We have the health care reform that the insurance and drug industries allowed us to have, and the financial regulation reform that the Wall Street banks bought and paid for. Even the consumer protection agency that Elizabeth Warren pushed and probably will not head is housed in the Federal Reserve, with major control in the hands of Bernanke and his successors. Meanwhile, nothing has been done on climate, and nothing has been done on immigration, both of which will require an honest rethinking of the "free trade" agreements and outsourcing. If we won't modify or back away from free trade, we cannot rebuild the middle class. The Senate cannot even extend unemployment benefits in a timely fashion. The regulatory agencies are in bed, as usual, with the industries they are charged with overseeing. The term banana republic jumps to mind.

Corporations have been with us since they were brought into being five centuries ago by the elites of Europe as a tool to use in confiscating the riches of the New World. They serve to pool capital by short-circuiting the public ethics attached to money. Shareholders are able to get returns from investing their money in return for reduced risk and very little in the way of legal liability.

Liability limits are the core of the problem with corporations. Liability is the way our society assigns responsibility for behavior. Separating morality from money results in Wall Street. And Wall Street fosters rot in our common culture.

The diagnosis is easy. Seeing the way out of the dilemma is not. It is difficult, not to say impossible, for a twenty-first century person to imagine what kind of economic system might have been built in these five centuries if certain fundamentals had not fallen into place for developing the corporate ethic. And if, as seems likely, we cannot change the Supreme Court's ruling about elections, how are we going to change the chartering of corporations?

On the other hand, change does happen. It is in fact continuous, and that we cannot see how to undo five centuries of history in order to organize our common lives differently does not mean that five centuries of history are not going to be undone or extensively modified. While I cannot see a solution, I can see a few possibilities that might lead in the right direction, toward setting some forces loose in the guts of the machine that may eventually tear it apart.

Most of these observations spring from thinking about whatever intentional communities we have in our midst. For farmers, the Amish, whose Christianity includes and informs their economic lives, are the most obvious teachers. They have their problems, and it won't do to minimize them. But they quite simply hold it to be wrong to allow an economic decision to destroy neighboring or neighborliness. That sensibility alone blows through our current moral and economic swamp like a sharp north wind in October. How can we make useful to us some of what the Amish have figured out? Here are a few thoughts, all of them conservative without being right wing.

- Gather close. The corporate structure has no interest in our wellbeing. Monsanto plans to own all the genetics for the earth's food supply. It has an effective enabler on the Supreme Court in Clarence Thomas. Other corporations are quietly cornering the earth's supply of useable water. This is an emergency, and we need to wake up. We shall have to look out for ourselves. Our first step is to bring our necessities close to us so that we can control them. Food needs to be locally sourced. Farming and gardening skills are essential. We need to know someone who can work with solar heating. We need to learn to patch our own pants. Someone who can construct clothes is a valuable community asset. Backyard mechanics and handymen are invaluable. None of this is cute or quaint. These things are vital, and we ignore them at our peril. Diversified funds such as 401ks or other investment in the stock market of extra or retirement money is suspect, and should be gotten out of. Our own communities need whatever money we have to invest. And we need to be regularly exercising real responsibility for our own behavior, including the use of our money. Not to do so is simply to keep saluting the same old delusions that got us here in the first place.

- Get smaller. A human-scale economy is a small-scale one. It is layered deep with smaller enterprises in areas such as food, shelter and fuel where we dare not fail. Such an approach delivers us from the illusion that we are safe because some big concern or some big machine is doing it for us, instead encouraging us to rely upon our own competence. And it teaches us that our economic safety rests on the success and goodwill of our neighbor. If my crop fails, his may not. This combination of self-reliance and cooperation is the best human response to the vagaries of life on earth, and in fact is the only safety net that can be counted upon.

- Go slower. Get some perspective. Speedboats and motorcycles and cul-de-sacs in the country are not important; food is. If it takes an hour or two to harvest the food and prepare the meal, let it. Then invite your neighbor to share. Perhaps he will bring some of his wine or beer. He may know a story you have yet to hear. Offer to help him build his chicken coop if he will help you figure out how to get your house to quit leaking whenever it rains. Elevate quality, and de-emphasize quantity. Get off the computer. Throw away the phone. Calm down. Think, and act.

HOW WE MIGHT LIVE

ARCHER

Visualize us standing beside an archer who is in the act of drawing back her bow, arrow notched to the string. At the very instant she releases the arrow, we are going to blow a very small puff of air at it from the side. The puff is so small, the archer will not feel it on her hands or face. The effort required to release the air toward the arrow is so slight that a bystander is not apt to notice it happening. But it is enough to change the course of the arrow ever so slightly.

This is the idea of the one-degree deflection.

If the archer is able to send the arrow a thousand feet into the distance, and we have managed to deflect the arrow by even one degree, it will land many feet wide of the archer's intended target, and toward the direction we are trying to encourage.

The archer is the national economy/culture, the arrow is the passage of time, and the puffer of air is, of course, anyone with the audacity to think that his or her life can make a difference.

This approach to change is necessary and protective of our sanity, for we live and farm in a powerful national and nationalizing economy that will not take kindly to any kind of real change, and that has immense power to block change. Much of this power inheres in the wants, desires, and thoughts of our own minds, so that we tend to enable this powerful economic structure while it vacuums the wealth out of our communities and the satisfaction from our lives.

I would not say that this slow and incremental change over a long period of time is the only change possible. Huge and relatively fast change happens frequently: think of the dissolution of the Soviet Union, or the end of apartheid in South Africa. Imagine what change will come from the passing of the peak in oil production, or some of the likely results of foolish government policies such as free trade and trade deficits, as well as our attempted military control of the entire globe.

But the slow change is change that each of us can drive. If we have a wholesome view of ourselves, such change will be beneficial to many people. It is less apt to be violent and more likely to be stable and permanent, rather than just being an excuse for the next war. If we have accustomed ourselves to searching out the needs and desires of our hearts and minds, and have learned to judge them and to put the good and necessary ones on the road to becoming reality in our own and our children's lives, we will be better equipped to cope with the change that huge events are going to force upon us, and possibly even turn some of that change to our advantage.

And we are fortunate, because many of us have some kind of ownership and management interest in a business that is biological — part of the great beating heart of the earth itself. Therefore we have a tool that is potentially more powerful than any held by any hand in Washington, D.C.

How do we use it? I think we start by taking up, one by one, the common, mundane things we deal with every day as we manage our farms and businesses, and examine them for meaning and opportunity. We must, for instance, regularly deal with labor. Generally this is our own labor, but also, sometimes and for some of us, that of others. Labor is related to physical work; in fact physical work before the advent of modern machines and technology would have been the entire definition of labor.

The idea of physical work has, in my lifetime, endured a complete change in meaning and significance. It was honorable in 1955; in 2007 it is not, having been replaced in our minds with the idea of games and the passing interest of the outdoor "sports enthusiast." A farm boy in 1955 would swell with pride at having first kept up with his elders at stacking hay; a farm boy in 2007 expects to "work" in a suit while occasionally "working out" in a weight and exercise room.

This compartmentalizing of work in our minds is not so much useful to us rural people as it is a very real marketing opportunity for the economy surrounding us. Because our bodies are less healthy, we are a market for the health sales companies. Since we will not walk, we can be sold gadgets to ride. What we will not put our hands on must be handled by a machine sold to us by a farm technology company. If we soon develop a horror of having to exert hands to control the machine we bought to do the work we didn't want to do, we can be persuaded to buy a computerized control for that purpose. The entire oversold biotech industry, from the bag of manufactured seed to the needle full of growth hormone, is a triumph of technology over labor and management.

It is important to realize that we have been sold this concept. It is one of the foundations of our entire economy, which functions to delete or at least outsource all labor so that profits may return exclusively to capital. But when we rural businesses outsource labor, we delete ourselves and shrink our communities. This is not something we will ever hear from the farm financial advisors. The disdain we have been taught for physical work translates into a knee-jerk tendency to always choose technology over labor in the management of our farms.

Richard Levins, formerly an economist with the University of Minnesota, uses four Schedule F-based measurements for farm sustainability. One is the amount paid out for labor, both in terms of the bottom-line return to operator labor and management, and the categories for hired labor and custom work hired. These labor costs are put into a ratio with the total expense. Levins concludes that for this measurement, the higher the ratio of labor to the total, the more sustainable the farm. He is right.

Actually getting there is a little complex and involves a route with which we are not entirely comfortable. The deck seems stacked against any move in this direction. This is where it gets discouraging. But remember the one-percent deflection.

For example, our farm has changed and grown to the point where we have two families living and working on just 320 acres, a situation unheard of here on the prairie. In addition, we are part of the cause for employment of several meat cutters and part of a trucking firm, all local. This works in large part because we have marketed the concept of local support to the buyers of our products.

But this also means we forgo some labor saving technologies. The flush toilet system of livestock production is not an option for us: The livestock housing and handling systems we certify to our customers dictate that we use straw, and therefore pitchforks. We keep a small-bale baler because that is the best way to feed calves, and our farrowing systems are based upon straw and grass.

None of this is the most "economical" method of livestock farming, but it is what people seem to want in their meat purchases, and it better fits the choices we want to make in terms of where we invest our money. Nothing is sadder than the farmer who keeps a brand new tractor, but who cannot afford decent shoes for the kids or a family vacation. We do not keep Sunday-go-to-meeting pickups, and we do not trade machinery when we are lean on income for family living.

We do keep a skid-loader that is in no danger of being disposed of anytime soon. It helps us cope with this huddle of obsolete livestock buildings. But we also have in the tool inventory a half-dozen useable pitchforks that go where the Bobcat cannot. I am instructing my grandkids in the use of these, so that when they get to the situation where fuel for the Bobcat gets to be a problem, they will understand that there is always another way.

Work that kids can do is a way of binding them to the farm so that they can see farming as an honorable occupation that uses mind and body both. It helps them imagine their lives proceeding on this land and among these people. They must be mentored in this by a respected adult who is not afraid to sometimes use his body instead of the latest technology.

If we are to continue any kind of life here in the country, we must at least demonstrate to our young this difference between us and the culture. Physical work is not necessarily drudgery. When it is done endlessly, hopelessly and alone, it approaches slavery. But when the point of it is readily apparent — when the doing of it produces a satisfaction regarding the health of the animals and the land, and when it is often done in the company of others — it can be an expression of joy, and a celebration of the life force within us.

The importance of dealing with technology cannot be underestimated if we are to get at the root of what bedevils us in rural America. People who think physical work is or ought to be beneath them, loneliness and alienation from the world around us including nature, broken relationships, lack of willingness to take a risk, general apathy, even the horrendous rates of murder, abuse and suicide in our nation — all are related at least in part to our unwillingness or inability to get technology into perspective. The powers that be are picking us off one at a time, and technology is one of their handiest tools because it can put us by ourselves and make us vulnerable. Let me give a few examples.

We use a computer in our business. We bought it at one of the super-duper "max" stores. When it locked up and quit two days short of warranty expiration, the super duper max store's troubleshooters tried to fix it over the phone. When that failed, they told us they would give us a box. We would need to drive the twenty miles to get the free box at their store. Then we could carefully box up the computer, drive the twenty miles to the bulk mailing place near their store, mail it, and then wait three weeks while they tried to fix it.

I have a phone in the house to which this computer is hooked for e-mail and word-processing purposes. The line got full of static and then went dead a few months ago. We were sure it was the line at fault, and so called the telephone company. The service man came out, looked the situation over, and said it was the phone. I didn't want to drive back to the super-duper store for another dose of long distance "service," so I told the service man, who I have known since we were kids, to bring a phone out when he got this direction again. This he did. It still works. The store wouldn't give us our money back on the defective phone, offering instead a credit that I don't want to use because I don't want any more of their stuff.

Technology has changed the way we make judgments. It was my pleasure to make presentations a few years ago about our farm and business on a four-stop tour in another part of the country with a woman who is one of the best graziers I know. The conversation got to overwintering at one point, and she told me she planned to try using horsepower to drag the big bales into position for feeding. Then she asked me not to mention it for fear the audiences wouldn't take her seriously if she were known to use horses.

Notice the lack of any human contact in the first example, other than the chance to pass a few words with the shipping clerk at the bulk mailing place. In the second, I opted for human contact with the telephone serviceman. It is important to note that it is only because of my somewhat advanced age that I even know that telephone service men have to do with — guess what? — telephones! And the third is simply a statement of how far we have come in a poorly considered direction that one of our brightest would question her own preference for biology over diesel fuel.

This is no diatribe against technology. But if we think about the difference between technology as a tool and technology as a master, we will begin to see choice as an option. In places this choice is, or should be, easy. A thirteen-year old boy who can't be talked to at home or in school because his nose is buried in the Game Boy is not a happy, well-adjusted kid. A farmer who must use a four-wheeler in the pasture in the daytime, and then a treadmill in the house in the evening to keep his heart in shape, is being silly. One who goes into deep hock to buy a GPS system to steer his tractor deserves to go broke trying to pay for it, as he has given away to the tech companies what little was left of his occupation.

But it is not always so easy. Assuming that we can learn to clearly see the difference between using and being used by technology, there still are going to be a lot of gadgets that could be argued either way. And in any case, the argument leaves out the human factor, which is where I started. The human factor is critical. Considering just electronics, we can see the line from radio to television, then computers to interactive internet and cellular phone technology, and how that line leads directly away from human interaction. It is possible to listen to radio in a group and talk at the same time, less so with television. At computers the human group drops off, replaced by semi-imaginary internet "friends." With interactive internet, even those friends have been replaced digitally with a machine that reads your responses and gives you what you want. A major complaint from adolescent girls is that adolescent boys are more interested in internet porn than in the girls. Maybe that's what ails my boars!

Bringing our humanity back into our thinking is going to be difficult, because it will feel like swimming upstream against progress. As my late friend Paul Gruchow observed, we have spent several generations preferring our neighbor's land to our neighbor, leaving us, as Leviticus warned, alone in the midst of the land. Any of us left in farming are at least somewhat guilty of that.

But there is sometimes a second chance, and it may be that history will offer us just that with the shortages of fuel and the deteriorating economic conditions. The question is, will we choose for people — for neighbors and family — if we get a do-over? Can we drive change toward that possibility by making better choices right now? Are we up to choosing people over other good things, such as production? There are no easy answers, but we should be catching on by now that for people to be whole and healthy, they must be needed.

Value-added businesses closely connected with farms and farm products have the potential to increase opportunity in rural areas. As an example, our business uses a local, state-inspected processor to cut and prepare the meats for store sales in the Twin Cities. Meat cutters employed by this rural processor are doing the job of several in-store meat counter employees in metropolitan areas. In developing this business, we have not only employed a local butcher, but gotten that shop to do work it otherwise would not be doing, thereby replacing employment in the urban area. Does Tyson Foods' accountant live in a rural place? Ours does, as well as the graphic designer who does our advertising, the print shop that prints our labels, and all of our top management (us).

This aspect of farm value-added businesses is significant, but it will not occur accidentally. There is no shortage of such businesses that operate with some or all of the most desirable work done in the urban areas, and where the top local management is the plant manager. Ethanol is an example of this, as is plant just up the road that generates electricity from turkey manure. Large confinement livestock production is prone to this model as well: think Smithfield Foods. Our labels are produced locally and our graphic design is rural in origin because our goals make it happen, not because there is any inevitability to it in this age of instant communication.

Here is where the one-percent deflection comes in. Our label business does not ensure the success of the small printer that does them. Though not a large business, the rural trucker we use for meat hauling would have thin soup indeed if he tried to live on just our contribution. Our accountant does business with many other individuals and businesses. We are not the sole customer of our meat processor. But in each of these cases we contribute to the success of the business, and we do so deliberately. It is more expensive to haul meat than live animals. But if we hauled the hogs to a processor in or near the Twin Cities and arranged the distribution from there, we would be supporting urban business and employment, and thus their civil infrastructure (schools, roads, services) rather than ours. The distinction is important to us because we have children and grandchildren here, and we need schools, stores and roads. We think it is largely up to us as rural businesses to provide them.

We would not give away the business to accomplish this. The difference in trucking the two products is significant, but not huge. Yet the investment keeps faith with our community. If our contribution helps make a rural business vital and viable, it is likely that other businesses and people will become customers as well, and a good thing we have helped to start may take root and grow, benefiting all.

Some of the jobs springing from a farm value-added business will be ordinary entry-level employment, and some will be more the professional category. It is important that we as rural folks not be talking down either type. We have some regrettable tendencies along those lines that will not be helpful in the building of a new rural economy.

We need more people on our farms, and some of these will be employees. Over the years we farmers have been impressed enough with ourselves as owners, with all the risk-taking and personal responsibility that involves, that we tend to project a kind of disdain for wage labor, using such terms as "farmhand" and "flunky." But a farm that wishes to produce and market livestock in a manner acceptable to a certain group of consumers is going to find it necessary to find, employ or cooperate with people who can relate to and handle livestock in non-confinement situations. These are not professional people, but they are few in number, and we need what they can provide. We need to act accordingly.

Similarly, we need professional people and sales staff. These folks must be respected for what they are and can do, and treated accordingly. For too many years we farmers have had it tough enough economically that we compensate by looking down on everyone else, thus propping up our own sense of value. Now, with the opportunities that are presenting themselves, we need to accept our own value. In doing this, we should find it easier to see the value in others. I never respected a salesman until I actually had to try to do it myself. Sales is hard, hard work, and without the sales staff the business falters and fails.

It is difficult to generate much excitement around the idea that economic improvement for our very rural areas is best accomplished by the way we make decisions, one farm and one business at a time. It seems so pokey, so slow. But it is good to remember that this progress is real, and that the gains, small as they may be, are apt to be longer lasting than those produced by the elaborate shell game by which various companies hold rural municipalities hostage for property tax giveaways and write-offs in exchange for locating there. These businesses generally don't outlast the buildings built to house them, and few ever generate enough local benefits to repay the lost tax income.

We should learn to think that the only help we ever will get is that which we give ourselves.

The powerful national, and increasingly international, economy is pretty good at posing as our friend. They work at it. Walmart has our neighbor's eighty-year old father saying hello and pushing a cart into our hands as we walk through the door. Monsanto commiserates with us as "dumb" urbanites and consumers criticize our use of Posilac and Roundup Ready. The tractor company thinks our life ought to be made easier by purchase of their product, and the computer industry shakes its electronic head and clucks sympathetically over the idea of us getting dirt under our fingernails.

Everyone hails us as "salt of the earth" and "the original environmentalist" while helping themselves to the lion's share of our profits. Meanwhile, we drive further for parts, for supplies and food, for school for the kids. We are living the modern rural version of an old philosophical chestnut about the falling tree in the forest. If a barn burns and no one is there to see it go, does it really burn?

This whole argument is going to lead to the drawing of some lines. There will be aspects of our lives where it is pretty hard to do that, where we are going to fall on opposite sides of a decision. But there are some easy areas. We have evolved to the point where we no longer stoop to pick up a penny, but think nothing of driving twenty miles to save a nickel. We need to question that tendency, which I found in full display at the farm store where I asked for Epsom salts, having a calf that needed its foot soaked. This store is part of a small Upper Midwest chain where I have gotten pretty used to shopping. These stores tend to close when a Walmart, Home Depot or Lowe's move into town.

The fellow I asked about the Epsom salts said he didn't have any, and suggested I run down to the Walmart about a mile distant where I could get a four-pound box "real cheap." I said we would pick it up instead at the grocery store next door where we were headed anyhow. He told me they probably had it, but warned that I would pay more. I bought the box in the grocery store for $2.39.

Someone needs his head examined in this story, and it is not me. The idea of doing business with friends has some very difficult aspects, but this is easy stuff. At a minimum we need to get these things right.

We need to internalize a set of sorting questions to run through mentally as we do our business. These questions have to do with who benefits from the transaction, how much of the benefit is local, and how much undesirable stuff (ecological, social, agricultural) cannot be separated from the thing being purchased.

For instance, one of my own sorting questions has to do with whether I want my money to be making dreams come true in Arkansas or New Jersey or Saudi Arabia or Texas. The answer is "no, I do not," but that doesn't mean I don't buy petroleum, insurance or too many cheap Chinese imports. What my voice accomplishes by inserting that question into my brain each time I buy one of these things is to make me edgy and uneasy about the decision. This uneasiness is going to keep me looking for alternatives, and will predispose me to seeing alternatives when they show up, as well as creating a few myself.

I am a patriot of western Minnesota. These prairies where I have spent my life and done my work are the landscape that is for me both home and homeland.

I realize I have used several words here that are conventionally loaded with both sentiment and violence, so let me be clear. I think one can only be patriotic toward that part of Creation he is familiar with and in which he has made himself at home. It is just this being at home that enables anyone to assume that others might feel the same way toward their homes. In a similar vein, patriotism toward the nation can only be built on the patriotism felt for the part of it that is right outside each of our doors. We cannot care for our nation without caring for our part of it.

Thus, for each of us it makes a difference where the goods and services come from. I can sit here in Chinese jeans typing on a computer made in Taiwan, invented in California and transported with fuel pumped from the ground in Texas. The economy has arranged things as a method of collecting maximum wealth into as few pockets as possible.

But the other economy, the one right outside my door, suffers and dies for lack of useful and meaningful work to do. As long as this is so, it is a marker for what we have yet to learn and the size of the changes yet to be made.

Many of the harder decisions have to do with technology because we are so accustomed to it, and because so much of it is imported from outside our neighborhoods. I have been on a number of grazing farms, and I hope to visit many more. What strikes me always is that the farms can be quite different, but the impulse is the same. That impulse is to make a living out of not fighting nature so much. Some farms have most of the latest technology, particularly in the animal handling and service areas, and some are old-fashioned to the point of being operated mostly by hand. These, of course, require less in the way of size or gross income.

I will not speak against technology, but I will say this: Rural areas have suffered from technology's tendency to replace humans for the last half century, at least. Therefore, I think one of our questions of technology needs to be about what we plan to do with the labor that the technology replaces. This is the Japanese style of thought. What can we do with people who have been replaced by technology, and how can we do it to retain more economic power in our local economies, rather than less? This question, should we be able to sort out some positive answers, holds more promise for our rural places than any of the national economy's bill of goods.

Do we have an impact in the world? It seems evident to me that we do. The depression we are so plagued with in our culture cloaks from us this most important understanding. But we know it; we tell it to each other in our stories. Consider the Christmas movie "It's a Wonderful Life" that is broadcast each year. Jimmy Stewart, in the lead role as a small-town banker with a commitment to helping ordinary folks buy homes and start businesses, gets into trouble because his forgetful uncle loses a large deposit. He falls into depression and is about to throw himself off a river bridge when an angel (who looks a lot like W.C. Fields) shows him a vision of what the town would look like if he hadn't spent his life as he did. It is one long strip mall complete with stores, traffic, people climbing over top of each other, saloons, brothels and more.

Stewart, his depression lifted by what he saw in the vision, goes home to pick up the pieces of his life. The movie gets too sappy for my taste here, but the point has been made. We do make a difference, and we know it. If we pay attention, we see this by how a thought, a saying or an act — helpful or harmful — reverberates in the behavior of the folks around us. That we often do not keep this knowledge of our human lives foremost in our minds, where it can maintain some control over our activities, is probably due as much as anything to the impact of a national/international economy that wants us reduced to nothing but a bundle of desires with a credit card.

The argument I am making here is self-evident to some and completely opaque to others. The difference has to do with where we see ourselves in the cosmic scheme. Do we get our satisfaction out of a day's productive work, or the latest electronic trinket? Does developing even a very partial understanding of our farm and its place in Creation seem like a useful life's work, or does a career consist of a new pickup every other year? Can we ever, in any circumstance, see the use and the need of work from which we will never benefit? Can we imagine our own lives coming to us from the distant past, and extending from us into the far future? There are seventy-year olds who plant trees every season here on the prairie, but by and large they are not the same seventy-year olds who are shopping for a new car before the last one is worn out. It is a matter of the level of spirituality, of sensing ourselves as living in eternity as well as in time.

If we do see our place in the universal order, and thus have no need of the argument I just made, then we can start considering applications. How, for instance, do I use for good my life's impact on the world? And we immediately come up against the question of "intentionality." Is it enough that I have a good heart and mean well? Maybe this is a job best left to pastor or priest? How about the certified smart guy at the university, or the politician?

I would argue that this is far too important to be left to the experts. It is for us who are entrusted with land and businesses, as well as the dependence of children and grandchildren. I think we need to plan. We have an obligation.

Anyone farming grassland soils in the Midwest ought to know that in the century since the breaking plows turned over these tough prairie sods, we have lost half of what we were given. (Or took from the Indians. Put your own interpretation on it.) Heavy, black topsoils that measured sixteen inches and two-feet in depth in settlement times now show clay subsoil on the high spots in far too many places. This is an incredible waste.

And it is especially tragic from the point of view of a grazier, since the farmers who did it had a vision and a view of the future in mind, but it was the wrong one, and it did not reach far enough in time. They saw the prairie as a potential Europe under complete cultivation in small, carefully rotated fields. These farmers evidently never considered the impact of thunderstorm rains and constant wind on the exposed soil. I see no evidence that any of them, before they took up that plow, considered what the Indians and the bison had going here.

We must plan. We have an obligation, those of us entrusted with land. And as Holistic Management teaches, we must try to reach seven generations, 150 years, into the future with our planning. History's mistakes show us that range is necessary. We know that if we do plan, much of that plan will be knocked awry by changing circumstances, by carelessness in government, greed in the economy, or by some other disaster our own carelessness might already have set loose upon the world. We know that we do not have the reach for this.

The point is not that we can control the future. Rather, it is that the distant vision will help govern our day-to-day management by serving as an early warning sign whenever we are headed in the wrong direction. If that prairie settler a century ago had held a dream and a vision for seven generations that included more perennial plants to ensure a healthy landscape, he may have had a second thought about destroying the sod he found here. Similarly for us, if our vision for the seventh generation includes more people on the land and more local wealth, as ours does for our farm, then we will tend to make decisions in a way that helps bring that about.

Seventh-generation planning seems beyond human capability. Perhaps it should be called seventh generation dreaming. But the numberless practical efforts that go into achieving the plan are not beyond us. Planning provides for us a measuring tool and an incentive that we need for the very human-scale task of living a decent life on earth.